# LIGHT SCIENCE FOR LEISURE HOURS.

*A SERIES OF FAMILIAR ESSAYS ON SCIENTIFIC SUBJECTS, NATURAL PHENOMENA, ETC.*

BY

RICHARD A. PROCTOR, B. A. CAMB., F. R. A. S.

AUTHOR OF
"THE SUN," "OTHER WORLDS THAN OURS," "SATURN," ETC.

"I bear you witness as ye bear to me,
Time, day, night, sun, stars, life, death, air, sea, earth."
SWINBURNE.

NEW YORK:
D. APPLETON AND COMPANY,
549 & 551 BROADWAY.
1871.

# PREFACE.

The Essays in the present volume have been selected from my contributions to serial literature during the past three or four years. Although I have for some time been urged to publish such a volume, I think I should not have ventured to do so but for the kindness with which my "Other Worlds" and "The Sun" have been received, both by the press and the public.

In preparing these Essays, my chief object has been to present scientific truths in a light and readable form—clearly and simply, but with an exact adherence to the facts as I see them. I have followed—here and always—the rule of trying to explain my meaning precisely as I should wish others to explain, to myself, matters with which I was unfamiliar. Hence I have avoided that excessive simplicity which some seem to consider absolutely essential in scientific essays intended for general perusal, but which is often even more perplexing than a too technical style. The chief rule I have followed, in order to make my descriptions clear, has been to endeavor to make each sentence bear one meaning, and one only. Speaking as a reader, and especially as a reader of scientific books, I venture to express an earnest wish that this simple rule were never infringed, even to meet the requirements of style.

It will hardly be necessary to mention that several of the shorter Essays are rather intended to amuse than to instruct.

The Essay on the influence which marriage has been supposed to exert on the death-rate is the one referred to by Mr. Darwin at page 176 (vol. i.) of his "Descent of Man."

This and the other Essays from the *Daily News* are selected from a large number of articles which I wrote in the years 1868–'70. It was by my kind friend Mr. Walker, formerly editor of the *Daily News*, that I was first urged to collect my Essays into a volume. I have to thank the proprietors and the present editor of the *Daily News*, and the proprietors and editors of the other journals from which the present series has been selected, for freely according me permission to reprint these Essays.

RICHARD A. PROCTOR.

LONDON, *May*, 1871.

# CONTENTS.

# LIGHT SCIENCE FOR LEISURE HOURS.

## *STRANGE DISCOVERIES RESPECTING THE AURORA.*

One of the most mysterious and beautiful of Nature's manifestations promises soon to disclose its secret. The brilliant streamers of colored light which wave at certain seasons over the heavens have long since been recognized as among the most singular and impressive of all the phenomena which the skies present to our view. There is something surpassingly beautiful in the appearance of the true "auroral curtain." Fringed with colored streamers, it waves to and fro as though shaken by some unseen hand. Then from end to end there pass a succession of undulations, the folds of the curtain interwrapping and forming a series of graceful curves. Suddenly, and as by magic, there succeeds a perfect stillness, as though the unseen power which had been displaying the varied beauties of the auroral curtain were resting for a moment. But even while the motion of the curtain is stilled we see its light mysteriously waxing and waning. Then as we gaze, fresh waves of disturbance traverse the magic canopy. Startling coruscations add splendor to the scene, while the noble span of the auroral arch from which the waving curtain seems to depend, gives a grandeur to the spectacle which no words can adequately describe. Gradually, however, the celestial fires which have illuminated the gorgeous arch seem to die out. The luminous zone breaks up. The scene of the display becomes covered with scattered streaks and patches of ashen gray light, which hang like clouds over the northern heavens. Then these in turn disappear, and nothing remains of the brilliant spectacle but a dark smoke-like segment on the horizon.

Such is the aurora as seen in arctic or antarctic regions, where the phenomenon appears in its fullest beauty. Even in our own latitudes, however, strikingly beautiful auroral displays may sometimes be witnessed. Yet those who have seen the spectacle presented near the true home of the aurora, recognize in other auroras a want of the fulness and splendor of color which form the most striking features of the arctic and antarctic auroral curtains.

Hitherto the nature of the aurora has been a mystery to men of science; nor, indeed, does the discovery we are about to describe throw even now full light on the character of the phenomenon. That discovery, however, affords promise of a speedy solution of the perplexing problems presented by auroral displays; and in itself, it is so full of interest and so suggestive, that our physicists already recognize it as one of the most important which have been made in recent times.

A few brief words in explanation of the progress which had been effected in the study of auroral phenomena, will serve to render the interest and importance of the discovery we have to describe more apparent.

Let it be premised, then, that physicists had long since recognized in the aurora a phenomenon of more than local, of more even than terrestrial, significance. They had learned to associate it with relations which affect the whole planetary scheme. Let us inquire how this had come about.

So long as men merely studied the appearances presented by the aurora, so long in fact as they merely regarded the phenomenon as a local display, they could form no adequate conception of its importance. The circumstance which first revealed something of the true character of the aurora was one which seemed to promise little.

Arago was engaged in watching from day to day, and from year to year, the vibrations of the magnetic needle in the Paris Observatory. He traced the slow progress of the needle to its extreme westerly variation, and watched its course as it began to retrace its way toward the true north. He discovered the minute vibration which the needle makes each day across its mean position. He noticed that this vibration is variable in extent; and so he was led to watch it more closely. Thus he had occasion to observe more attentively than had yet been done the sudden irregularities which occasionally characterize the daily movements of the needle.

All this seems to have nothing to do with the auroral streamers; but we now reach the important discovery which rewarded Arago's patient watchfulness.

In January, 1819, he published a statement to the effect that the sudden changes of the magnetic needle are often associated with the occurrence of an aurora. I give the statement in his own words, as translated by General Sabine: "Auroras ought to be placed in the first rank among the causes which sometimes disturb the regular march of the diurnal changes of the magnetic needle. These do not, even in summer, exceed a quarter of a degree, but when an aurora appears, the magnetic needle is often seen to move in a few instants over several degrees." "During an aurora," he adds, "one often sees in the northern region of the heavens luminous streamers of different colors shoot from all points of the horizon. The point in the sky to which these streamers converge is precisely the point to which a magnetized needle suspended by its centre of gravity directs itself. . . . . . It has, moreover, been shown that the concentric circular segments, almost similar in form to the rainbow, which are usually seen previous to the appearance of the luminous streamers, have their two extremities resting on two parts of the horizon which are equally distant from the direction toward which the needle turns; and the summit of each arc lies exactly in that direction. *From all this it appears, incontestably, that there is an intimate connection between the causes of auroras and those of terrestrial magnetism.*"

This strange hypothesis was, at first, much opposed by scientific men. Among others the late Sir David Brewster pointed out a variety of objections, some of which appeared at first sight of great force. Thus, he remarked that magnetic disturbances of the most remarkable character have often been observed when no aurora has been visible; and he noticed certain peculiarities in the auroras observed near the polar regions, which did not seem to accord with Arago's view.

But gradually it was found that physicists had mistaken the character of the auroral display. It appeared that the magnetic needle not only swayed responsively to auroras observable in the immediate neighborhood, but to auroras in progress hundreds or even thousands of miles away. Nay, as inquiry progressed, it was discovered that the needles in our northern observatories are swayed by influences associated even with the occurrence of auroras around the southern polar regions.

In fact, not only have the difficulties pointed out (very properly, it need hardly be remarked) by Sir David Brewster been wholly removed; but it has been found that a much closer bond of sympathy exists between the magnetized needle and the auroral streamers than even Arago had supposed. It is not merely the case that while an auroral display is in progress the needle is subject to unusual disturbance, but the movements of the needle are actually synchronous with the waving movements of the mysterious streamers. An aurora may be in progress in the north of Europe, or even in Asia or America, and as the colored banners wave to and fro, the tiny needle, watched by patient observers at Greenwich or Paris, will respond to every phase of the display.

And I may notice in passing that two very interesting conclusions follow from this peculiarity: First, every magnetic needle over the whole earth must be simultaneously disturbed; and, secondly, the auroral streamers which wave across the skies of one country, must move synchronously with those which are visible in the skies of another country, even though thousands of miles may separate the two regions.

But I must pass on to consider further the circumstances which give interest and significance to the strange discovery which is the subject of this paper.

Could we only associate auroras with terrestrial magnetism, we should still have done much to enhance the interest which the beautiful phenomenon is calculated to excite. But when once this association has been established, others of even greater interest are brought into recognition. For terrestrial magnetism has been clearly shown to be influenced directly by the action of the sun. The needle in its daily vibration follows the sun, not indeed through a complete revolution, but as far as the influence of other forces will permit. This has been abundantly confirmed, and is a fact of extreme importance in the theory of terrestrial magnetism. Wherever the sun may be, either on the visible heavens or on that half of the celestial sphere which is at the moment beneath the horizon, the end of the needle nearest to the sun makes an effort (so to speak) to point more directly toward the great ruling centre of the planetary scheme. Seeing, then, that the daily vibration of the needle is thus caused, we recognize the fact that the disturbances of the daily vibration may be referred to some peculiarity of the solar action.

It was not, therefore, so surprising as many have supposed, that the increase and diminution of these disturbances, in a period of about eleven years, should be found to correspond with the increase and diminution of the number of solar spots in a period of equal length.

We already begin to see, then, that auroras are associated in some mysterious way with the action of the solar rays. The phenomenon which had been looked on for so many ages as a mere spectacle, caused perhaps by some process in the upper regions of the air, of a simply local character, has been brought into the range of planetary phenomena. As surely as the brilliant planets which deck the nocturnal skies are illuminated by the same orb which gives us our days and seasons, so are they subject to the same mysterious influence which causes the northern banners to wave resplendently over the starlit depths of heaven. Nay, it is even probable that every flicker and coruscation of our auroral displays corresponds with similar manifestations upon every planet which travels round the sun. It becomes, then, a question of exceeding interest to inquire what is the nature of the mysterious apparition which from time to time illuminates our skies. We have learned something of the laws according to which the aurora appears; but what is its true nature? What sort of light is that which illuminates the heavens? Is there some process of combustion going on in the upper regions of our atmosphere? Or are the auroral streamers electric or phosphorescent? Or, lastly, is the light simply solar light reflected from some substance which exists at an enormous elevation above the earth?

All these views have from time to time found supporters among scientific men. It need hardly be said that what we now know of the association between auroral action and some form of solar disturbance, would at once enable us to reject some of these hypotheses. But we need not discuss the subject from this point of view; because a mode of research has recently been rendered available which at once answers our inquiries as to the general character of any kind of light. I proceed to consider the application of this method to the light from the auroral streamers.

The spectroscope, or, as we may term the instrument, the "light-sifter," tells us of what nature an object which is a source of light may be. If the object is a luminous solid or liquid, the instrument converts its light into a rainbow-colored streak. If the object is a luminous vapor, its light is converted into a few bright lines. And, lastly, if the object is a luminous solid or liquid shining through any vapors, the rainbow-colored streak again makes its appearance, but it is now crossed by dark lines, corresponding to the vapors which surround the object and absorb a portion of its light.

But I must not omit to notice two circumstances which render the interpretation of a spectrum somewhat less simple than it would otherwise be.

In the first place, if an object is shining by reflected light, its spectrum is precisely similar to that of the object whose light illuminates it. Thus we cannot pronounce positively as to the nature of an object merely from the appearance of its spectrum, unless we are quite certain that the object is self-luminous. For example, we observe the solar spectrum to be a rainbow-colored streak crossed by a multitude of dark lines, and we conclude accordingly that the sun is an incandescent globe shining through a complex vaporous atmosphere. We feel no doubt on this point, because we are absolutely certain that the sun is self-luminous. Again, we observe the spectrum of the moon to be exactly similar to the solar spectrum, only, of course, much less brilliant. And here also we feel no doubt in interpreting the result. We know, certainly, that the moon is not self-luminous, and therefore we conclude with the utmost certainty that the light we receive from her is simply reflected solar light. So far all is clear. But now take the case of an object like a comet, which may or may not be self-luminous. If we find that a comet's spectrum resembles the sun's—and this is not altogether an hypothetical case, for a portion of the light of every comet yet examined does in reality give a rainbow-colored streak resembling the solar spectrum—we cannot form, in that case, any such positive conclusion. The comet may be a self-luminous body, but, on the other hand, its light may be due merely to the reflection of the solar beams. Accordingly, we find that our spectroscopists always accompany the record of such an observation with an expression of doubt as to the real nature of the object which is the source of light.

Secondly, when an electric spark flashes through any vapor, its light gives a spectrum which indicates the nature, not only of the vapor through which the spark has passed, but of the substances between which the spark has travelled. Thus, if we cause an electric flash to pass between iron points through common air, we see in the spectrum the numerous bright lines which form the spectrum of iron, and in addition we see the bright lines belonging to the gases which form our atmosphere.

Both the considerations above discussed are of the utmost importance in studying the subject of the auroral light as analyzed by the spectroscope, because there are many difficulties in forming a general opinion as to the nature of the auroral light, while there are circumstances which would lead us to anticipate that the light is electric.

We notice also in passing that we owe to the German physicist Angström a large share of the researches on which the above results respecting the spectrum of the electric spark are founded. The reader will presently see why we have brought Angström's name prominently forward in connection with the interesting branch of spectroscopic analysis just referred to. If the discovery we are approaching had been effected by a tyro in the use of the spectroscope, doubts might very reasonably have been entertained respecting the exactness of the observations on which the discovery rests.

It was suggested many years ago, long indeed before the true powers of spectroscopic analysis had been revealed, that perhaps if the light of the aurora were analyzed by the prism, evidence could be obtained of its electric nature. The eminent meteorologist Dové remarked, for instance, that "the peculiarities presented by the electric light are so marked that it appears easy to decide definitely by prismatic analysis, whether the light of the aurora is or is not electric." Singularly enough, however, the first proof that the auroral light is of an electric nature was derived from a very different mode of inquiry. Dr. Robinson, of Armagh, discovered in 1858 (a year before Kirchhoff's recognition of the powers of spectroscopic analysis) that the light of the aurora possesses in a peculiar degree a property termed fluorescence, which is a recognized and characteristic property of the light produced by electrical discharges. "These effects," he remarks of the appearances presented by the auroral light under the tests he applied, "were so strong in relation to the actual intensity of the light, that they appear to afford an additional evidence of the electric origin of the phenomenon."

Passing over this ingenious application of one of

the most singular and interesting properties of light, we find that the earliest determination of the real nature of the auroral light—or rather of its spectrum—was that effected by Angström. This observer took advantage of the occurrence of a brilliant aurora in the winter of 1867–'68 to analyze the spectrum of the colored streamers. *A single bright line only was seen!* Otto Struve, an eminent Russian astronomer, shortly afterward made confirmatory observations. At the meeting of the Royal Astronomical Society in June, 1868, Mr. Huggins, F. R. S., thus described Struve's results: "In a letter, M. Otto Struve has informed me that he has had two good opportunities of observing the spectrum of the aurora borealis. The spectrum consists of one line, and the light is therefore monochromatic. The line falls near the margin of the yellow and green portions of the spectrum. . . . This shows that the monochromatic light is greenish, which surprised me; but General Sabine tells me that in his polar expeditions he has frequently seen the aurora tinged with green, and this appearance corresponds with the position of the line seen by M. Struve."

The general import of this observation there is no mistaking. It teaches us that the light of the aurora is due to luminous vapor, and we may conclude, with every appearance of probability, that the luminosity of the vapor is due to the passage of electric discharges through it. It is, however, possible that the position of the bright line may be due to the character of the particles between which the discharge takes place.

But the view we are to take must depend upon the position of the line. Here a difficulty presents itself. There is no known terrestrial element whose spectrum has a bright line precisely in the position of the line in the auroral spectrum. And mere proximity has no significance whatever in spectroscopic analysis. Two elements differing as much from each other in character as iron and hydrogen may have lines so closely approximating in position that only the most powerful spectroscope can indicate the difference. So that when Angström remarks that the bright line he has seen lies slightly to the left of a well-known group of lines belonging to the metal calcium (the principal ingredient of common chalk), we are by no means to infer that he supposes the substance which causes the presence of the bright line has any resemblance to that element. Until we can find an element which has a bright line in its spectrum absolutely coincident with the bright line detected by Angström in the spectrum of the aurora,* all speculation as to the real nature of the vapor in which the auroral electric discharge takes place, or of the substance between which the spark travels, is altogether precluded.

But interesting as the discovery undoubtedly is,

* Other green lines have since been discovered in the auroral spectrum; and occasionally a red line is seen.

we have now to deal with one of a yet more interesting character.

Most of my readers have doubtless heard of the zodiacal light, and many of them have perhaps seen that mysterious radiance, pointing obliquely upward from the western horizon soon after sunset in the spring months, or in autumn shortly before sunrise, above the eastern horizon. The light, as its name indeed implies, lies upon that region of the heavens along which the planets travel. Accordingly, astronomers have associated it with the planetary orbits, and have come to look on it as formed by the light reflected from a multitude of minute bodies travelling around the sun within the orbit of our earth.

Yet it had long been recognized that there are difficulties in the way of this theory. Passing over those which depend on the position of the zodiacal light upon the heavens, there are difficulties connected with the appearance of the object. For example, its light has often been observed to flicker or coruscate in a manner which it seemed difficult to ascribe to the motions of our own atmosphere. Then again there have been seasons when the zodiacal light has shone with unusual intensity for months together, and there is nothing in the received theory which can account for such a peculiarity. Lastly, there is the strange circumstance recorded by Baron Humboldt that the zodiacal light is often invisible when night first sets in, and then suddenly appears with full splendor; a phenomenon which is utterly inexplicable if the received theory be accepted. The whole account of the phenomenon, as given by Baron Humboldt, is so interesting, and for my present purpose so significant, that I give it at full length:

"In the tropical climate of South America," he remarks, "the variable strength of the light of the zodiacal gleam struck me at times with utter amazement. As I there passed the beautiful nights, in the open air, on the banks of rivers, and in the grassy plains for several months together, I had opportunities of observing the phenomenon with attention. When the zodiacal light was at its very brightest, it sometimes happened that but a few mintues afterward it became notably weakened, and then it suddenly gleamed up again with its former brilliancy. In particular instances, I believed that I remarked—not any thing of a ruddy tinge, or an interior arched obscuration, or an emission of sparks, such as Mairan describes, but—a kind of unsteadiness and flickering of the light."

Despite these and similar observations, very little doubt had been felt by astronomers that the zodiacal light really indicates the presence of minute bodies travelling in more or less eccentric paths round the sun. And it was confidently expected that whenever a spectroscope of sufficient delicacy to analyze the faint light of the zodiacal gleam was applied to that purpose, the resulting spectrum would be merely a very faint reproduction of the solar spectrum.

Recently, however, the zodiacal light has been analyzed by Angström, with a result altogether unexpected, and at present almost unintelligible. *Its spectrum exhibits a bright line, and this bright line is the same that is seen in the spectrum of the aurora borealis!*

How are we to understand this most surprising result? Remembering that the aurora is undoubtedly a terrestrial light, whencesoever it derives its luminosity—in other words, that the electric discharges, however excited, really take place in the upper regions of our own atmosphere, while as certainly the zodiacal light is an extra-terrestrial phenomenon—the observed phenomenon becomes one of the most perplexing discoveries ever made by man. That it will before long be interpreted we have no doubt whatever; nor do we doubt that the interpretation will involve the explanation of a whole series of phenomena which have lately perplexed astronomers. Recalling the association between auroras and terrestrial magnetism, and that between terrestrial magnetism and the solar spots, and remembering further that our physicists have recently detected well-marked signs that the planets in their courses influence the sun's atmosphere and generate his spots in some manner as yet unexplained, we see that the one fact wanting to explain Angström's discovery is undoubtedly not an isolated fact, but must be associated in the most intimate manner with a variety of important cosmical relations. To speculate as to the nature of the as yet undiscovered interpretation of Angström's researches would at present be an idle task, perhaps. But one feature of the solar scheme with which we cannot doubt that it will be found to be associated, must be mentioned before we conclude.

Of all the phenomena presented to the contemplation of astronomers, the tails of comets are undoubtedly the most perplexing. Their rapid formation, their swift motions (if, indeed, we could believe that their changes of position are due to a real transmission of their material substance), and the enormous variety of configuration and of structure which they present to our contemplation, render them not merely amazing, but altogether unintelligible.

Now, there is one feature of comets' tails which has long since attracted attention, and will remind the reader of the peculiarities common to the zodiacal and the auroral light. We refer to the sudden changes of brilliancy, the flickerings or coruscations, and the instantaneous lengthening and shortening of these mysterious appendages. Olbers spoke of "explosions and pulsations which in a few seconds went trembling through the whole length of a comet's tail, with the effect now of lengthening, now of abridging it by several degrees." And the eminent mathematician

Euler was led by the observation of similar appearances to put forward the theory "*that there is a great affinity between these tails, the zodiacal light, and the aurora borealis.*" The late Admiral Smyth, commenting on this opinion of Euler's, remarks that "most reasoners seem now to consider comets' tails as consisting of electric matter;" adding that "this would account for the undulations and other appearances which have been noticed, as, for instance, that extraordinary one seen by M. Chladni in the comet of 1811, when certain undulatory ebullitions rushed from the nucleus to the end of the tail, a distance of more than ten millions of miles, in two or three seconds of time." To this we may add the somewhat bizarre theory suggested by Sir John Herschel, that the matter forming the zodiacal light is "loaded, perhaps, with the actual materials of the tails of millions of comets, which have been stripped of these appendages in the course of successive passages round the immediate neighborhood of the sun."

Now, hitherto no comet with a sufficiently brilliant tail for spectroscopic analysis has appeared since Kirchhoff's invention of that mode of research. Already our physicists had been looking forward anxiously for the appearance of such a comet as Donati's or Halley's. But Angström's recent discovery, and the evidence which seems to associate the tails of comets with the auroral and zodiacal lights, render our spectroscopists doubly anxious to submit a comet's tail to spectroscopic analysis. It is far from being unlikely that three long-vexed questions—the nature of the aurora, and that of the zodiacal light, and that of comets' tails—will receive their solution simultaneously.

I had scarcely completed the above pages when news was brought from America that the spectrum of the sun's corona, as seen during the recent total solar eclipse, exhibited the same bright lines as the aurora. The fact that auroral *lines* are mentioned will at once be noticed; but it is to be remarked that the two faint lines which have been lately seen in the auroral spectrum correspond to but a very small portion of the light we receive from the northern streamers. In the spectrum of the corona the same three lines appear, but their relative brightness is different. The brightest line of the auroral spectrum is faint in the spectrum of the corona, while the latter exhibits a bright line where the former has a faint one.

News has also been received that a comparison of the photographs of the eclipse proves the corona, or at any rate its brightest part, to belong to the sun.

Lastly, it has been found that the peculiar phosphorescent light sometimes visible all over the sky at night gives the same spectrum (very faint, of course) as the aurora and the zodiacal light.

It is impossible not to recognize the fact that these discoveries point to relations of the utmost importance. The teachings of the spectroscope are too certain to be mistaken. When it shows us such and such lines bright or dark, we may conclude, without fear of being misled, that such and such substances are emitting or absorbing light. What we learn certainly, therefore, from the facts above stated, is this, that substances of the same sort emit the light of the aurora, of the zodiacal gleam, of the sun's corona, and of the phosphorescence which illuminates at times the nocturnal skies. We may conclude, but not so certainly, that the manner in which the light is emitted is also the same in each case. We know certainly that the auroral light is excited by the solar action. We know certainly that it is associated with the earth's magnetism. The opinion, then, which we should form of the source to which the other lights are due is tolerably obvious. So long as electricity was merely used as a convenient way of accounting for any perplexing phenomenon, it was impossible to accept explanations of cosmical peculiarities as due to electrical action. But when once we have reason —as in the case of the aurora we undoubtedly have—to associate electricity with any particular form of luminosity, we seem clearly justified in extending the explanation to the same form of luminosity wherever it may appear.

I believe that the key to the whole series of phenomena dealt with above lies in the existence of myriads of meteoric bodies travelling separately or in systems round the sun. They are consumed in thousands daily by our own atmosphere; they probably pour in countless millions upon the solar atmosphere; and from what we know of their numbers in our own neighborhood, and of the probability of their being infinitely more numerous in the neighborhood of the sun, we have excellent reasons for believing that to them principally is due the appearance of the zodiacal light and the solar corona.

(From *Fraser's Magazine*, February, 1870.)

---

## *THE EARTH A MAGNET.*

There is a very prevalent but erroneous opinion that the magnetic needle points to the north. We remember well how we discovered in our boyhood that the needle does *not* point to the north, for the discovery was impressed upon us in a very unpleasant manner. We had purchased a pocket-compass, and were very anxious—not, indeed, to test the instrument, since we placed implicit reliance upon its indications—but to make use of it as a guide across unknown regions. Not many miles from where we lived lay Cobham Wood, no very extensive forest certainly, but large enough to lose one's self in. Thither, accordingly, we proceeded with three school-fellows. When we had lost ourselves, we gleefully called the compass into action, and made from the wood in a direction which we supposed would lead us home. We travelled on with full confidence in our pocket-guide; at each turning we consulted it in an artistic manner, carefully poising it and waiting till its vibrations ceased. But when we had travelled some two or three miles without seeing any house or road that we recognized, matters assumed a less cheerful aspect. We were unwilling to compromise our dignity as "explorers" by asking the way—a proceeding which no precedent in the history of our favorite travellers allowed us to think of. But evening came on, and with it a summer thunderstorm. We were getting thoroughly tired out, and the *hæc olim meminisse juvabit* with which we had been comforting ourselves began to lose its force. When at length we yielded, we learned that we had gone many miles out of our road, and we did not reach home till several hours after dark. How it fared with our school-fellows we know not, but a result overtook ourselves personally, for which there is no precedent, so far as we are aware, in the records of exploring expeditions. Also the offending compass was confiscated by justly indignant parents, so that for a long while the cause of our troubles was a mystery to us. We now know that instead of pointing due north, the compass pointed more than 20° toward the west, or nearly to the quarter called by sailors north-northwest. No wonder, therefore, that we went astray when we followed a guide so untrustworthy.

The peculiarity that the magnet needle does not, in general, point to the north, is the first of a series of peculiarities which we now propose briefly to describe. The irregularity is called by sailors the needle's *variation*, but the term more commonly used by scientific men is the *declination* of the needle. It was probably discovered a long time ago, for 800 years before our era the Chinese applied the magnet's directive force to guide them in journeying over the great Asiatic plains; and they must soon have detected so marked a peculiarity. Instead of a ship's compass, they made use of a magnetic car, on the front of which a floating needle carried as mall figure, whose outstretched arm pointed southward. We have no record, however, of their discovery of the declination, and know only that they were acquainted with it in the twelfth century. The declination was discovered, independently, by European observers in the thirteenth century.

As we travel from place to place, the declination of the needle is found to vary. Christopher Columbus was the first to detect this. He discovered it on the 13th of September, 1492, during his first voyage, and when he was six hundred miles from Ferro, the most westerly of the Canary Islands. He found that the declination, which was toward the east in Europe, passed to the west, and increased continually as he travelled westward.

But here we see the first trace of a yet more singu-

.ar peculiarity. We have said that at present the declination is toward the west in Europe. In Columbus's time it was toward the east. Thus we learn that the declination varies with the progress of time, as well as with change of place.

The genius of modern science is a weighing and a measuring one. Men are not satisfied nowadays with knowing that a peculiarity exists; they seek to determine its extent, how far it is variable—whether from time to time or from place to place, and so on. Now the results of such inquiries applied to the magnetic declination have proved exceedingly interesting.

We find, first, that the world may be divided into two unequal portions, over one of which the needle has a westerly, and over the other an easterly, declination. Along the boundary-line, of course, the needle points due north. England is situated in the region of westerly magnets. This region includes all Europe, except the northeastern parts of Russia; Turkey, Arabia, and the whole of Africa; the greater part of the Indian Ocean, and the western parts of Australia; nearly the whole of the Atlantic Ocean; Greenland, the eastern parts of Canada, and a small slice from the northeastern part of Brazil. All these form one region of westerly declination; but, singularly enough, there lies in the very heart of the remaining and larger region of easterly magnets an oval space of a contrary character. This space includes the Japanese Islands, Mantchooria, and the eastern parts of China. It is very noteworthy also, that in the westerly region the declination is much greater than in the easterly. Over the whole of Asia, for instance, the needle points almost due north. On the contrary, in the north of Greenland and of Baffin's Bay, the magnetic needle points due west; while still farther to the north (a little westerly), we find the needle pointing with its north end directly toward the south.

In the presence of these peculiarities, it would be pleasant to speculate. We might imagine the existence of powerfully magnetic *veins* in the earth's solid mass, coercing the magnetic needle from a full obedience to the true polar summons. Or the comparative effects of oceans and of continents might be called into play. But, unfortunately for all this, we have to reconcile views founded on *fixed* relations presented by the earth with the process of *change* indicated above. Let us consider the declination in England alone.

In the fifteenth century there was an easterly declination. This gradually diminished, so that in about the year 1657 the needle pointed due north. After this the needle pointed toward the west, and continually more and more, so that scientific men, having had experience only of a continual shifting of the needle in one direction, began to form the opinion that this change would continue, so that the needle would pass, through northwest and west, to the south. In fact, it was imagined that the motion of the needle would resemble that of the hands of a watch, only in a reversed direction. But before long observant men detected a gradual diminution in the needle's westerly motion. Arago, the distinguished French astronomer and physicist, was the first (we believe) to point out that "the progressive movement of the magnetic needle toward the west appeared to have become continually slower of late years" (he wrote in 1814), "which seemed to indicate that after some little time longer it might become retrograde." Three years later, namely, on the 10th of February, 1817, Arago asserted definitively that the retrograde movement of the magnetic needle had commenced to be perceptible. Colonel Beaufoy at first oppugned Arago's conclusion, for he found from observations made in London during the years 1817–1819, that the westerly motion still continued. But he had omitted to take notice of one very simple fact, viz., that London and Paris are two different places. A few years later the retrograde motion became perceptible at London also, and it has now been established by the observations of forty years. It appears, from a careful comparison of Beaufoy's observations, that the needle reached the limit of its western digression (at Greenwich) in March, 1819, at which time the declination was very nearly 25°. In Paris, on the contrary, the needle had reached its greatest western digression (about 22½°) in 1814. It is rather singular that, although at Paris the retrograde motion thus presented itself five years earlier than in London, the needle pointed due north at Paris six years later than in London, viz., in 1663. Perhaps the greater amplitude of the needle's London digression may explain this peculiarity.

"It was already sufficiently difficult," says Arago, "to imagine what could be the kind of change in the constitution of the globe which could act during one hundred and fifty-three years in gradually transferring the direction of the magnetic needle from due north to 23° west of north. We see that it is now necessary to explain, moreover, how it has happened that this gradual change has ceased, and has given place to a return toward the preceding state of the globe." "How is it," he pertinently asks, "that the directive action of the globe, which clearly must result from the action of molecules of which the globe is composed, can be thus variable, while the number, position, and temperature of these molecules, and, as far as we know, all their other physical properties, remain constant?"

But we have considered only a single region of the earth's surface. Arago's opinion will seem still more just when we examine the change which has taken place in what we may term the "magnetic aspect" of the whole globe. The line which separates the region of westerly magnets from the region of easterly magnets now runs, as we have said, across Canada and eastern Brazil in one hemisphere, and across Russia, Asiatic Turkey, the Indian Ocean, and West Australia, in the other; besides having an outlying oval to the east of the Asiatic continent. Now these lines have swept round a part of the globe's circuit in a most singular manner since 1600. They have varied alike in direction and complexity. The Siberian oval, now distinct, was in 1787 merely a loop of the eastern line of no declination. The oval appears now to be continually diminishing, and will one day probably disappear.

We find here presented to us a phenomenon as mysterious, as astonishing, and as worthy of careful study, as any embraced in the wide domains of science. But other peculiarities await our notice.

If a magnetic needle of suitable length be carefully poised on a fine point, or, better, be suspended from a silk thread without torsion, it will be found to exhibit each day two small, but clearly perceptible, oscillations. M. Arago, from a careful series of observations, deduced the following results:

At about eleven at night, the north end of the needle begins to move from west to east, and having reached its greatest easterly excursion at about a quarter-past eight in the morning, returns toward the west to attain its greatest westerly excursion at a quarter-past one. It then moves again to the east, and having reached its greatest easterly excursion at half-past eight in the evening, returns to the west, and attains its greatest westerly excursion at eleven, as at starting.

Of course, these excursions take place on either side of the mean position of the needle, and as the excursions are small, never exceeding the fifth part of a degree, while the mean position of the needle lies some 20° to the west of north, it is clear that the excursions are only nominally eastern and western, the needle pointing, throughout, far to the west.

Now if we remember that the north end of the needle is that farthest from the sun, it will be easy to trace in M. Arago's results a sort of effort on the part of the needle to turn toward the sun—not merely when that luminary is above the horizon, but during his nocturnal path also.

We are prepared, therefore, to expect that a variation, having an annual period, shall appear, on a close observation of our suspended needle. Such a variation has been long since recognized. It is found that in the summer of both hemispheres, the daily variation is exaggerated, while in winter it is diminished.

But besides the divergence of a magnetized needle from the north pole, there is a divergence from the horizontal position which must now claim our attention. If a non-magnetic needle be carefully suspended so as to rest horizontally, and be then magnetized, it will be found no longer to preserve that position. The northern end *dips* very sensibly. This happens in our hemisphere. In the southern, it is the southern end which dips. It is clear, therefore, that if we travel from one hemisphere to the other, we must find the northern dip of the needle gradually diminishing, till at some point near the equator the needle is horizontal; and as we pass thence to southern regions, a gradually increasing southern inclination is presented. This has been found to be the case, and the position of the line along which there is no inclination (called the *magnetic equator*) has been traced around the globe. It is not coincident with the earth's equator, but crosses that circle at an angle of twelve degrees, passing from north to south of the equator in long. 3° west of Greenwich, and from south to north in long. 187° east of Greenwich. The form of the line is not exactly that of a great circle, but presents here and there (and especially where it crosses the Atlantic) perceptible excursions from such a figure.

At two points on the earth's globe the needle will rest in a vertical position. These are the magnetic poles of the earth. The northern magnetic pole was reached by Sir J. G. Ross, and lies in 70° N. lat. and 263° E. long., that is, to the north of the American Continent, and not very far from Boothia Gulf. One of the objects with which Ross set out on his celebrated expedition to the Antarctic Seas was the discovery, if possible, of the southern magnetic pole. In this he was not successful. Twice he was in hopes of attaining his object, but each time he was stopped by a barrier of land. He approached so near, however, to the pole, that the needle was inclined at an angle of nearly ninety degrees to the horizon, and he was able to assign to the southern pole a position in 75° S. lat., 154° E. long. It is not probable, we should imagine, that either pole is fixed, since we shall now see that the inclination, like the declination of the magnetic needle, is variable from time to time, as well as from place to place; and in particular, the magnetic equator is apparently subjected to a slow but uniform process of change.

Arago tells us that the inclination of the needle at Paris has been observed to diminish year by year since 1671. At that time the inclination was no less than 75°; in other words, the needle was inclined only 15° to the vertical. In 1791 the inclination was less than 71°. In 1831 it was less than 68°. In like manner, the inclination at London has been observed to diminish, from 72° in 1786 to 70° in 1804, and thence to 68° at the present time.

It might be anticipated from such changes as these that the magnetic equator would be found to be changing in position. Nay, we can even guess in which way it must be changing. For, since the inclination is diminishing at London and Paris, the magnetic equator must be approaching these places, and this (in the present position of the curve) can only happen by a gradual shifting of the magnetic equator from east to west along the true equator. This motion has been found to be really taking place. It is supposed that the movement is accompanied by a change of form; but more observations are necessary to establish this interesting point.

Can it be doubted that while these changes are taking place, the magnetic poles also are slowly shifting round the true pole? Must not the northern pole, for instance, be farther from Paris now that the needle is inclined more than 23° from the vertical, than in 1671, when the inclination was only 15°? It appears obvious that this must be so, and we deduce the interesting conclusion that each of the magnetic poles is rotating around the earth's axis.

But there is another peculiarity of the needle which is as noteworthy as any of those we have spoken about. We refer to the intensity of the magnetic action—the energy with which the needle seeks its position of rest. This is not only variable from place to place, but from time to time, and is further subject to sudden changes of a very singular character.

It might be expected that where the dip is greater, the directive energy of the magnet would be proportionately great. And this is found to be approximately the case. Accordingly, the magnetic equator is very nearly coincident with the "equator of least intensity," but not exactly. As we approach the magnetic poles we find a more considerable divergence, so that instead of there being a northern pole of greatest intensity nearly coincident with the northern magnetic pole, which we have seen lies to the north of the American Continent, there are *two* northern poles, one in Siberia nearly at the point where the river Lena crosses the Arctic circle, the other not so far to the north—only a few degrees north, in fact, of Lake Superior. In the south, in like manner, there are also two poles, one on the Antarctic circle, about 130° E. long., in Adelie Island, the other not yet precisely determined, but supposed to lie on about the 240th degree of longitude, and south of the Antarctic circle. Singularly enough, there is a line of lower intensity running right round the earth along the valleys of the two great oceans, "passing through Behring's Straits and bisecting the Pacific, on one side of the globe, and passing out of the Arctic Sea by Spitzbergen and down the Atlantic, on the other."

General Sabine discovered that the intensity of the magnetic action varies during the course of the year. It is greatest in December and January *in both hemispheres.* If the intensity had been greatest in winter, one would have been disposed to have assigned seasonal variation of temperature as the cause of the change. But as the epoch is the same for both hemispheres, we must seek another cause. Is there any astronomical element which seems to correspond with the law discovered by Sabine? There is one very important element. The position of the perihelion of the earth's orbit is such that the earth is nearest to the sun on about the 31st of December or the 1st of January. There seems nothing rashly speculative, then, in concluding that the sun exercises a magnetic influence on the earth, varying according to the distance of the earth from the sun. Nay, Sabine's results seem to point very distinctly to the law of variation. For, although the number of observations is not as yet very great, and the extreme delicacy of the variation renders the determination of its amount very difficult, enough has been done to show that in all probability the sun's influence varies according to the same law as gravity—that is, inversely as the square of the distance.

That the sun, the source of light and heat, and the great gravitating centre of the solar system, should exercise a magnetic influence upon the earth, and that this influence should vary according to the same law as gravity, or as the distribution of light and heat, will not appear perhaps very surprising. But the discovery by Sabine that *the moon* exercises a distinctly traceable effect upon the magnetic needle seems to us a very remarkable one. We receive very little light from the moon, much less (in comparison with the sun's light) than most persons would suppose, and we get absolutely no perceptible heat from her. Therefore it would seem rather to the influence of mass and proximity that the magnetic disturbances caused by the moon must be ascribed. But if the moon exercises an influence in this way, why should not the planets? We shall see that there is evidence of some such influence being exerted by these bodies.

More mysterious, if possible, than any of the facts we have discussed is the phenomenon of *magnetic storms.* The needle has been exhibiting for several weeks the most perfect uniformity of oscillation. Day after day, the careful microscopic observation of the needle's progress has revealed a steady swaying to and fro, such as may be seen in the masts of a stately ship at anchor on the scarce-heaving breast of ocean. Suddenly a change is noted; irregular jerking movements are perceptible, totally distinct from the regular periodic oscillations. A magnetic storm is in progress. But where is the centre of disturbance, and what are the limits of the storm? The answer is remarkable. If the jerking movements observed in places spread over very large regions of the earth—and in some well-authenticated cases over the whole earth—be compared with the local time, it is found that (allowance being made for difference of longitude) *they occur precisely at the same instant.* The magnetic vibrations thrill in one moment through the whole frame of our earth!

But a very singular circumstance is observed to characterize these magnetic storms. They are nearly always observed to be accompanied by the exhibition of the aurora in high latitudes, northern and southern. Probably they never happen without such a display; but numbers of auroras escape our notice. The converse proposition, however, *has* been established as a universal one. No great display of the aurora ever occurs without a strongly-marked magnetic storm.

Magnetic storms sometimes last for several hours or even days.

Remembering the influence which the sun has been found to exercise upon the magnetic needle, the question will naturally arise, Has the sun any thing to do with magnetic storms? We have clear evidence that he has.

On the 1st of September, 1859, Messrs. Carrington and Hodgson were observing the sun, one at Oxford and the other in London. Their scrutiny was directed to certain large spots which, at that time, marked the sun's face. Suddenly a bright light was seen by each observer to break out on the sun's surface and to travel, slowly in appearance, but in reality at the rate of about 7,000 miles in a minute, across a part of the solar disk. Now it was found afterward that the self-registering magnetic instruments at Kew had made at that very instant a strongly-marked jerk. It was learned that at that moment a magnetic storm prevailed at the West Indies, in South America, and in Australia. The signal-men in the telegraph-stations at

Washington and Philadelphia received strong electric shocks; the pen of Bain's telegraph was followed by a flame of fire; and in Norway the telegraphic machinery was set on fire. At night great auroras were seen in both hemispheres. It is impossible not to connect these startling magnetic indications with the remarkable appearance observed upon the sun's disk.

But there is other evidence. Magnetic storms prevail more commonly in some years than in others. In those years in which they occur most frequently, it is found that the ordinary oscillations of the magnetic needle are more extensive than usual. Now, when these peculiarities had been noticed for many years, it was found that there was an alternate and systematic increase and diminution in the intensity of magnetic action, and that the period of the variation was about eleven years. But at the same time, a diligent observer had been recording the appearance of the sun's face from day to day and from year to year. He had found that the solar spots are in some years more freely displayed than in others. And he had determined the period in which the spots are successively presented with maximum frequency to be about eleven years. On a comparison of the two sets of observations, it was found (and has now been placed beyond a doubt by many years of continued observation) that magnetic perturbations are most energetic when the sun is most spotted, and *vice versa*.

For so remarkable a phenomenon as this none but a cosmical cause can suffice. We can neither say that the spots cause the magnetic storms, nor that the magnetic storms cause the spots. We must seek for a cause producing at once both sets of phenomena. There is as yet no certainty in this matter, but it seems as if philosophers would soon be able to trace in the disturbing action of the planets upon the solar atmosphere the cause as well of the marked period of eleven years as of other less distinctly-marked periods which a diligent observation of solar phenomena is beginning to educe.

(From the *Cornhill Magazine*, June, 1868.)

---

## *OUR CHIEF TIMEPIECE LOSING TIME.*

A DISTINGUISHED French astronomer, author of one of the most fascinating works on popular astronomy that has hitherto appeared, remarks that a man would be looked upon as a maniac who should speak of the influence of Jupiter's moons upon the cotton-trade. Yet, as he proceeds to show, there is an easily-traced connection between the ideas which appear at first sight so incongruous. The link is found in the determination of celestial longitude.

Similarly, what would be thought of an astronomer who, regarding thoughtfully the stately motion of the sidereal system, as exhibited on a magnified, and therefore appreciable, scale by a powerful telescope, should speak of the connection between this movement and the intrinsic worth of a sovereign? The natural thought with most men would be that "too much learning" had made the astronomer mad. Yet, when we come to inquire closely into the question of a sovereign's intrinsic value, we find ourselves led to the diurnal motion of the stars, and that by no very intricate path. For, What is a sovereign? A coin containing so many grains of gold mixed with so many grains of alloy. A grain, we know, is the weight of such and such a volume of a certain standard substance—that is, so many cubic inches, or parts of a cubic inch, of that substance. But what is an inch? It is determined, we find, as a certain fraction of the length of a pendulum vibrating seconds in the latitude of London. A second, we know, is a certain portion of a mean solar day, and is practically determined by a reference to what is called a sidereal day—the interval, namely, between the successive passages by the same star of the celestial meridian of any fixed place. This interval is assumed to be constant, and it has indeed been described as the "one constant element" known to astronomers.

We find, then, that there is a connection, and a very important connection, between the motion of the stars and our measures, not merely of value, but of weight, length, volume, and time. In fact, our whole system of weights and measures is founded on the apparent diurnal motion of the sidereal system, that is, on the real diurnal rotation of the earth. We may look on the meridian-plane in which the great transit-telescope of the Greenwich Observatory is made to swing, as the gigantic hand of a mighty dial, a hand which, extending outward among the stars, traces out for us, by its motion among them, the exact progress of time, and so gives us the means of weighing, measuring, and valuing terrestrial objects with an exactitude which is at present *beyond* our wants.

The earth, then, is our "chief timepiece," and it is of the correctness of this giant clock that we are now to speak.

But how can we test a timepiece whose motions we select to regulate every other timepiece? If a man sets his watch every morning by the clock at Westminster, it is clearly impossible for him to test the accuracy of that clock by the motions of his watch. It would, indeed, be possible to detect any gross change of rate; but, for the purpose of illustration, I assume, what is indeed the case, that the clock is very accurate, and, therefore, that minute errors only are to be looked for even in long intervals of time. And, just as the watch set by a clock cannot be made use of to test the clock for small errors, so our best timepieces cannot be employed to detect slow variations, if any such exist, in the earth's rotation-period.

Sir William Herschel, who early saw the importance of the subject, suggested another method. Some of the planets rotate in such a manner, and bear such distinct marks upon their surface, that it is possible, by a series of observations extending over a long interval of time, to determine the length of their rotation-period within a second or two. Supposing their rotation uniform, we at once obtain an accurate measure of time. Supposing their rotation *not* uniform, we obtain—(1) a hint of the kind of change we are looking for; and (2), by the comparison of two or more planets, the means of guessing how the variation is to be distributed between the observed planets and our own earth.

Unfortunately, it turned out that Jupiter, one of the planets from which Herschel expected most, does *not* afford us exact information—his real surface being always veiled by his dense and vapor-laden atmosphere. Saturn, Venus, and Mercury, are similarly circumstanced, and are in other respects unfavorable objects for this sort of observation. Mars only, of all the planets, is really available. Distinctly marked (in telescopes of sufficient power) with continents and oceans, which are rarely concealed by vapors, this planet is in other respects fortunately situated. For it is certain that whatever variations may be taking place in planetary rotations must be due to external agencies. Now, Saturn and Jupiter have their satellites to influence (perhaps appreciably in long intervals of time) their rotation-movements. Venus and Mercury are near the sun, and are therefore in this respect worse off than the earth, whose rotation is in question. Mars, on the other hand, farther removed than we are from the sun, having also no moon, and being of small dimensions (a very important point, be it observed, since the tidal action of the sun depends on the dimensions of a planet), is likely to have a rotation-period all but absolutely constant.

Herschel was rather unfortunate in his observations of Mars. Having obtained a rough approximation from Mars' rotation in an interval of two days—this rough approximation being, as it happened, only thirty-seven seconds in excess of the true period—he proceeded to take three intervals of one month each. This should have given a much better value, but, as it happened, the mean of the values he obtained was forty-six seconds too great. He then took a period of two years, and being misled by the erroneous values he had already obtained, he *missed one rotation*, getting a value two minutes too great. Thirty years ago, two German astronomers, Beer and Mädler, tried the same problem, and taking a period of seven years, obtained a value which exceeds the true value by only one second. Another German, Kaiser, by combining more observations, obtained a value which is within one-fifteenth of a second of the true value. But a comparison of observations extending over 200 years has enabled the present writer to obtain a value which he considers to lie one-hundredth part of a second of the truth. This value for Mars' rotation-period is 24 hours 37 minutes 22·74 seconds.

Here, then, we have a result so accurate, that *at some future time*, it may serve to test the earth's rotation-period. We have compared the rotation-rate of our test-planet with the earth's rate during the past 200 years; and therefore, if the earth's rate vary by more than one-hundredth of a second in the next two or three hundred years, we shall—or, rather, our descendants will—begin to have some notion of the change at the end of that time.

But, in the mean time, mankind being impatient, and not willing to leave to a distant posterity any question which can possibly be answered *now*, astronomers have looked around them for information available at once on this interesting point. The search has not been in vain. In fact, we are able to announce, with an approach to positiveness, that our great terrestrial timepiece is actually *losing time*.

In our moon we have a neighbor which has long been in the habit of answering truthfully questions addressed to her by astronomers. Of old, she told Newton about gravitation, and when he doubted, and urged contradictory evidence offered—as men in his time supposed—by the earth, she set him on the right track, so that when in due time the evidence offered by the earth was corrected, Newton was prepared at once to accept and propound the noble theory which rendered his name illustrious. Again, men wished to learn the true shape of the earth, and went hither and thither measuring its globe; but the moon, meanwhile, told the astronomer who remained at home a truer tale. They sought to learn the earth's distance from the sun, and from this and that point they turned their telescopes on Venus in transit; but the moon has set them nearer the truth, and that not by a few miles, but by 3,000,000 or more. We shall see that she has had something to say about our great terrestrial timepiece.

One of the great charms of the science of astronomy is, that it enables men to *predict.* At such and such an hour, the astronomer is able to say, a celestial body will occupy such and such a point on the celestial sphere. You direct a telescope toward the point named, and lo! at the given instant the promised orb sweeps across the field of view. Each year there is issued a thick octavo volume crowded with such predictions, three or four years in advance of the events predicted; and these predictions are accepted with as little doubt by astronomers as if they were the records of past events.

But astronomers are not only able to predict—they can also trace back the paths of the celestial bodies, and say: "At such and such a long-past epoch, a given star or planet occupied such and such a position upon the celestial sphere." But how are they to verify such a statement? It is clear that, in general, they cannot do so. Those who are able to appreciate (or better, to make use of) the predictions of astronomy, will, indeed, very readily accord a full measure of confidence to calculations of past events. They know that astronomy is justly named the most exact of the sciences, and they can see that there is nothing, in the nature of things, to render retrospection more difficult than prevision. But there are hundreds who have no such experience of the exactness of modern astronomical methods—who have, on the contrary, a vague notion that modern astronomy is merely the successor of systems now exploded; perhaps even that it may one day have to make way in its turn for new methods. And if all other men were willing to accept the calculations of astronomers respecting long-past events, astronomers themselves would be less easily satisfied. Long experience has taught them that the detection of error is the most fruitful source of knowledge; therefore, wherever such a course is possible, they always gladly submit their calculations to the test of observation.

Now, looking backward into the far past, it is only here and there that we see records which afford means of comparison with modern calculations. The planets have swept on in their courses for ages with none to note them. Gradually, observant men began to notice and record the more remarkable phenomena. But such records, made with very insufficient instrumental means, have in general but little actual value. It has been found easy to confirm them without any special regard to accuracy of calculation.

But there is one class of phenomena which no inaccuracy of observation can very greatly affect. A total eclipse of the sun is an occurrence so remarkable, that (1) it can hardly take place without being recorded, and (2) a very rough record will suffice to determine the particular eclipse referred to. Long intervals elapse between successive total eclipses visible at the same place on the earth's surface; and even partial eclipses of noteworthy extent occur but seldom at any assigned place. Very early, therefore, in the history of modern astronomy, the suggestion was made, that eclipses recorded by ancient historians should be calculated retrospectively. An unexpected result rewarded the undertaking: it was found that ancient eclipses could not be fairly accounted for without assigning a slower motion to the moon in long-past ages than she has at present!

Here was a difficulty which long puzzled mathematicians. One after another was foiled by it. Halley, an English mathematician, had detected the difficulty, but no English mathematician was able to grapple with it. Contented with Newton's fame, they had suffered their Continental rivals to shoot far ahead in the course he had pointed out. But the best Continental mathematicians were defeated. In papers of acknowledged merit, adorned by a variety of new processes, and showing a deep insight into the question at issue, they yet arrived, one and all, at the same conclusion—failure.

Ninety years elapsed before the true explanation was offered by the great mathematician Laplace. A full exposition of his views would be out of place in such a paper as the present, but briefly, they amount to this:

The moon travels in her orbit, swayed chiefly by the earth's attraction. But the sun, though greatly more distant, yet, owing to the immensity of his mass, plays an important part in guiding our satellite. His influence tends to relieve the moon, in part, from the earth's sway. Thus she travels in a wider orbit, and with a slower motion, than she would have but for the sun's influence. Now, the earth is not at all times equally distant from the sun, and his influence upon the moon is accordingly variable. In winter, when the earth is nearest to the sun, his influence is greatest. The lunar month, accordingly (as any one may see by referring to an almanac), is longer in winter than in summer. This variation had long been recognized as the moon's "annual equation;" but Laplace was the first to point out that the variation is itself slowly varying. The earth's orbit is slowly changing in shape—becoming more and more nearly circular year by year. As the greater axis of her orbit is unchanging, it is clear that the actual extent of the orbit is slowly increasing. Thus, the moon is slightly released from the sun's influence year by year, and so brought more and more under the earth's influence. She travels, therefore, continually faster and faster; though the change is indeed but a very minute one—only to be detected in long intervals of time. Also the moon's *acceleration*, as the change is termed, is only temporary, and will in due time be replaced by an equally gradual retardation.

When Laplace had calculated the extent of the change due to the cause he had detected, and when it was found that ancient eclipses were now satisfactorily accounted for, it may well be believed that there was triumph in the mathematical camp. But this was not all. Other mathematicians attacked the same problem, and their results agreed so closely that all were convinced that the difficulty was thoroughly vanquished.

A very noteworthy result flowed from Laplace's calculations. Among other solutions which had been suggested, was the supposition (supported by no less an authority than Sir Isaac Newton, who lived to see the commencement of the long conflict maintained by mathematicians with this difficulty), that it is not the moon travelling more quickly, but our earth rotating more slowly, which causes the observed discrepancy. Now, it resulted from Laplace's labors—as he was the first to announce—that the period of the earth's rotation has not varied by one-tenth of a second per century in the last two thousand years. The question thus satisfactorily settled, as was supposed, was shelved for more than a quarter of a century. The result, also, which seemed to flow from the discussion—the constancy of the earth's rotation-movement—was accepted; and, as we have seen, our national system of measures was founded upon the assumed constancy of the day's duration.

But mathematicians were premature in their rejoicings. The question has been brought, by the labors of Professor Adams—co-discoverer with Leverrier of the distant Neptune—almost exactly to the point which it occupied a century ago. We are face to face with the very difficulties—somewhat modified in extent but not in character—which puzzled Halley, Euler, and Lagrange. It would be an injustice to the memory of Laplace to say that his labors were thrown away. The explanation offered by him is indeed a just one, but it is insufficient. Properly estimated, it removes only half the difficulty which had perplexed mathematicians. It would be quite impossible to present in brief space, and in form suited to these pages, the views propounded by Adams. What, for instance, would most of our readers learn if we were to tell them that, "when the variability of the eccentricity is taken into account, in integrating the differential equations involved in the problem of the lunar motions—that is, when the eccentricity is made a function of the time —non-periodic or secular terms appear in the expression for the moon's mean motion"—and so on? Let it suffice to say that Laplace had considered only the effect of the sun in diminishing the earth's *pull* on the moon, supposing that the slow variation in the sun's *direct* influence on the moon's motion in her orbit must be self-compensatory in long intervals of time. Adams has shown, on the contrary, that when this variation is closely examined, no such compensation is found to take place; and that the effect of this want of compensation is to diminish by more than one-half the effects due to the slow variation examined by Laplace.

These views gave rise at first to considerable controversy. Pontecoulant characterized Adams's processes as "analytical conjuring-tricks;" and Leverrier stood up gallantly in defence of Laplace. The contest swayed hither and thither for a while; but gradually the press of new arrivals on Adams's side began to prevail. One by one, his antagonists gave way; new processes have confirmed his results, figure for figure; and no doubt now exists, in the mind of any astronomer competent to judge, of the correctness of Adams's views.

But side by side with this inquiry, another had been in progress. A crowd of diligent laborers had been searching with close and rigid scrutiny into the circumstances attending ancient eclipses. A new light had been thrown upon this subject by the labors of modern travellers and historians. One remarkable instance of this may be cited. Mr. Layard has identified the site of Larissa with the modern Nimroud. Now, Xenophon relates that when Larissa was besieged by the Persians, an eclipse of the sun took place, so remarkable in its effects (and therefore undoubtedly total), that the Median defenders of the town threw down their arms and the city was accordingly captured. And Hansen had shown that a certain estimate of the moon's motion makes the eclipse which occurred on August 15, 310 B. C., not only *total*, but *central* at Nimroud. Some other remarkable eclipses—as the celebrated sunset eclipse (total) at Rome 399 B. C., the eclipse which enveloped the fleet of Agathocles as he escaped from Syracuse; the famous eclipse of Thales, which interrupted a battle between the Medes and Lydians; and even the partial eclipse which (probably) caused the "going back of the shadow upon the dial of Ahaz"—have all been accounted for satisfactorily by Hansen's estimate of the moon's motion; so, also, have nineteen lunar eclipses recorded in the Almagest.

This estimate of Hansen's, which accounts so satisfactorily for solar and lunar eclipses, makes the moon's rate of motion increase more than twice as fast as it should do according to the calculations of Adams. But before our readers run away with the notion that astronomers have here gone quite astray, it will be well to present, in a simple manner, the extreme minuteness of the discrepancy about which all the coil has been made.

Suppose that, just in front of our moon, a false moon exactly equal to ours in size and appearance (see *note* at the end of this paper) were to set off with a motion corresponding to the present motion of the moon, save only in one respect—namely, that the false moon's motion should not be subject to the change we are considering, termed *the acceleration.* Then one hundred years would elapse before our moon would fairly begin to show in advance. She would, in that time, have brought only one-one-hundred-and-fiftieth part of her breadth from behind the false moon. At the end of another century, she would have gained four times as much; at the end of a third, nine times as much: and so on. She would not fairly have cleared her own breadth in less than twelve hundred years. But the *whole* of this gain, minute as it is, is not left unaccounted for by our modern astronomical theories. *Half* the gain is explained, the other half remains to be interpreted; in other words, *the moon travels farther by about half her own breadth in twelve cen-*

*turies than she should do according to the lunar theory.*

But in this difficulty, small as it seems, we are not left wholly without resource. We are not only able to say that the discrepancy is probably due to a gradual retardation of the earth's rotation-movement, but we are able to place our finger on a very sufficient cause for such a retardation. One of the most firmly-established principles of modern science is this—that where *work is done*, force is, in some way or other, expended. The *doing of work* may show itself in a variety of ways—in the generation of heat, in the production of light, in the raising of weights, and so on; but in every case an equivalent force must be expended. If the brakes are applied to a train in motion, intense heat is generated in the substance of the brake; now, the force employed by the brakesman is *not* equivalent to the heat generated. Where, then, is the balance of force expended? We all know that the train's motion is retarded, and this loss of motion represents the requisite expenditure of force. Now, is there any process in Nature resembling, in however remote a degree, the application of a brake to check the earth's rotation? There is. The tidal wave, which sweeps, twice a day, round the earth, travels in a direction contrary to the earth's motion of rotation. That this wave "does work," no one can doubt who has watched its effects. The mere rise and fall in open ocean may not be strikingly indicative of "work done;" but when we see the behavior of the tidal wave in narrow channels, when we see heavily-laden ships swept steadily up our tidal rivers, we cannot but recognize the expenditure of force. Now, where does this force come from? Motion being the great "force-measurer," what motion *suffers* that the tides may *work?* We may securely reply, that the only motion which *can* supply the requisite force is the earth's motion of rotation. Therefore, it is no idle dream, but a matter of absolute certainty, that, though slowly, still very surely, our terrestrial globe is losing its rotation-movement.

Considered as a timepiece, what are the earth's errors? Suppose, for a moment, that the earth was *timed* and *rated* two thousand years ago, how much has she *lost*, and what is her "rate error?" She has lost in that interval nearly one hour and a quarter, and she is losing now at the rate of one second in twelve weeks. In other words, the length of a day is now more by about one-eighty-fourth part of a second than it was two thousand years ago. At this rate of change, our day would merge into a lunar month in the course of thirty-six thousand millions of years. But after a while, the change will take place more slowly, and some trillion or so of years will elapse before the full change is effected.

Distant, however, as is the epoch at which the

changes we have been considering will become effective, the subject appears to us to have an interest apart from the mere speculative consideration of the future physical condition of our globe. Instead of the recurrence of ever-varying, closely-intermingled cycles of fluctuation, we see, now for the first time, the evidence of cosmical decay—a decay which, in its slow progress, may be but the preparation for renewed genesis—but still, a decay which, so far as the races at present subsisting upon the earth are concerned, must be looked upon as finally and completely destructive.*

(From *Chambers's Journal*, October 12, 1867.)

---

## *ENCKE THE ASTRONOMER.*

Four years have passed since Encke died. Even those four years have witnessed notable changes in the aspect of the science he loved so well. But we must look back over more than fifty years, if we would form an estimate of the position of astronomy when Encke's most notable work was achieved. At Seeberge, under Lindenau, Encke had been perfecting himself in the

* In the *Quarterly Journal of Science* for October, 1866, a more detailed but somewhat less popular account of the subject of the above paper is presented. A few months earlier, a charmingly-written paper on the same subject, from the pen of Mr. J. M. Wilson, of Rugby, had appeared in the *Eagle*, a magazine written by and for members of St. John's College, Cambridge. Although my paper in the *Quarterly Journal of Science* was written quite independently of Mr. Wilson's (which, however, I had read), yet it chanced that in describing the same mathe-

higher branches of mathematical calculation. He took the difficult work of determining the orbital motions of newly-discovered comets under his special charge, and Dr. Bruhns tells us that every comet which was detected during Encke's stay at Seeberge was subjected to rigid scrutiny by the indefatigable mathematician. Before long a discovery of the utmost importance rewarded his persevering labors. Pons had detected on November 26, 1818, a comet of no very brilliant aspect, which was watched first at Marseilles, and then at Mannheim, until December 29th. Encke next took up

matical relations, and the same sequence of events, I here and there used language closely resembling his. I fear this led for a while to some misconception; but I was fortunately able to show in Mr. De la Rue's address to the Astronomical Society, on the same subject, passages yet more strikingly resembling some in Mr. Wilson's paper (written subsequently and quite independently). The fact would seem to be that, if two persons describe the same events, and deal with the same mathematical relations, it is almost certain that in more than one passage they will use somewhat similar expressions.

I was actually indebted to Mr. Wilson's paper for one illustration, however—that derived from the movements of a supposed artificial moon; and I think that, had his paper appeared in a magazine printed for general circulation, I should have referred to it. As it was, this seemed useless so far as the readers of the *Quarterly Journal of Science* were concerned. The circumstances of the case were, indeed, far from calling for a reference; while I had in a sense made the illustration my own by detecting an important miscalculation in the original (the amount of advance being either doubled or halved—I forget which). Had I referred to Mr. Wilson's paper, I must needs have mentioned this mistake; and it would have appeared as though I had no other purpose in making the reference.

I mention these matters to explain what I fear my esteemed fellow-collegian was disposed at the time to regard as either a wrong or a slight. Nothing was further from my intention than either.

the work, and tracked the comet until January 12th. Combining the observations made between December 22d and January 12th, he assigned to the body aparabolic orbit. But he was not satisfied with the accordance between this path and the observed motions of the body. When he attempted to account for the motions of the comet by means of an orbit of comparatively short period, he was struck by the resemblance between the path thus deduced and that of Comet I, 1805. Gradually the idea dawned upon him that a new era was opening for science. Hitherto the only periodical comets which had been discovered except Lexell's—the "lost comet"—had travelled in orbits extending far out into space beyond the paths of the most distant known planets. But now Encke saw reason to believe that he had to deal with a comet travelling within the orbit of Jupiter. On February 5th, he wrote to the eminent mathematician Gauss, pointing out the results of his inquiries, and saying that he only waited for the encouragement and authority of his former teacher to prosecute his researches to the end toward which they already seemed to point. Gauss, in reply, not only encouraged Encke to proceed, but counselled him as to the course he should pursue. The result we all know. Encke showed conclusively that the newly-discovered comet travels in a path of short period, and that it had already made its appearance several times in our neighborhood.

From the date of this discovery, Encke took high rank among the astronomers of Europe. His subsequent labors by no means fell short of the promise which this, his first notable achievement, had afforded. If he effected less as an astronomical observer than many of his contemporaries, he was surpassed by few as a manipulator of those abstruse formulæ by which the planetary perturbations are calculated. It was to the confidence engendered by this skill that we owe his celebrated discovery of the acceleration of the motion of the comet mentioned above. Assured that he had rightly estimated the disturbances to which the comet is subjected, he was able to pronounce confidently that some cause continually (though all but imperceptibly) impedes the passage of this body through space, and so—by one of those strange relations which the student of astronomy is familiar with—the continually retarded comet travels ever mores wiftly along a continually diminishing orbit.

Bruhns's Life of Encke is well worth reading, not only by those who are interested in Encke's fame and work as an astronomer, but by the general reader. Encke the man is presented to our view, as well as Encke the astronomer. With loving pains the pupil of the great astronomer handles the theme he has selected. The boyhood of Encke, his studies, his soldier-life in the great uprising against Napoleon in 1813, and his work at the Seeberge Observatory; his labors on comets and asteroids; his investigations of the transits of 1761 and 1769; his life as an academician, and as director of an important observatory; his orations at festival and funeral; and lastly, his illness and death, are described in these pages by one who held Encke in grateful remembrance as "teacher and master," and as "a fatherly friend."

Not the least interesting feature of the work is the correspondence introduced into its pages. We find Encke in communication with Humboldt, with Bessel and Struve, with Hansen, Olbers, and Argelander; with a host, in fine, of living as well as of departed men of science.

(From *Nature*, March 10, 1870.)

---

## *VENUS ON THE SUN'S FACE.*

More than a century ago scientific men were looking forward with eager interest to the passage of the planet Venus across the sun's face in 1769. The Royal Society judged the approaching event to be of such extreme importance to the science of astronomy, that they presented a memorial to King George III., requesting that a vessel might be fitted out, at government expense, to convey skilful observers to one of the stations which had been judged suitable for observing the phenomenon. The petition was complied with, and, after some difficulty as to the choice of a leader, the good ship "Endeavor," of 370 tons, was placed under the command of Captain Cook. The astronomical work intrusted to the expedition was completely successful; and thus it was held that England had satisfactorily discharged her part of the work of utilizing the rare phenomenon known as a transit of Venus.

A century passes, and science is again awaiting with interest the approach of one of these transits. But now her demands are enlarged. It is not one ship that is asked for, but the full cost and charge of several expeditions. And this time, also, science has been more careful in taking time by the forelock. The first hints of her requirements were heard some fourteen years ago, when the Astronomer-Royal began that process of laborious inquiry which a question of this sort necessarily demands. Gradually, her hints became more and more plain-spoken; insomuch that Mr. Airy—her mouth-piece in this case—stated definitely, a few months ago, what he thought science had a right to claim from England in this matter. When the claim came before our government, it was met with a liberality which was a pleasing surprise after Mr. Lowe's placid reference of scientific people to their own devices. The sum of ten thousand five hundred pounds has been granted to meet the cost of several important and well-appointed expeditions; and doubtless further material aid will be derived from the various government observatories.

And now let us inquire why so much interest is attached to a phenomenon which appears, at first sight, to be so insignificant. Transits, eclipses, and other phenomena of that nature, are continually occurring, without any particular interest being attached to them. The telescopist may see half a dozen such phenomena in the course of a night or two, by simply watching the satellites of Jupiter, or the passage of our moon over the stars. Even the great eclipse of 1868 did not attract so much interest as the coming transit of Venus, yet that eclipse had never been equalled in importance by any which has occurred in historic times, and hundreds of years must pass before such another happens, whereas transits of Venus are far from being so uncommon.

The fact is, that Venus gives us the best means we have of mastering a problem which is one of the most important within the whole range of the science of astronomy. We use the term important, of course, with reference to the scientific significance and interest of the problem. Practically, it matters little to us whether the sun is a million of miles or a thousand millions of miles from us. The subject must in any case be looked upon as an extra-parochial one. But science does occasionally attach immense interest to extra-parochial subjects. And this is neither unwise nor unreasonable, since we find implanted in our very nature—and not merely in the nature of scientific men—a quality which causes us to take interest in a variety of matters that do not in the least concern our personal interests. Nor is this quality, rightly considered, one of the least noble characteristics of the human race.

That the determination of the sun's distance is important, in an astronomical sense, will be seen at once when it is remembered that the ideas we form of the dimensions of the solar system are wholly dependent on our estimate of the sun's distance. Nor can we gauge the celestial depths with any feeling of assurance, unless we know the true length of that which is our sole measuring-rod. It is, in fact, our basis of measurement for the whole visible universe. In some respects, even if we knew the sun's distance exactly, it would still be an unsatisfactory gauge for the stellar depths. But that is the misfortune, not the fault of the astronomer, who must be content to use the measuring-rod which Nature gives him. All he can do is to find out as nearly as he can its true length.

When we come to consider how the astronomer is to determine this very element—the sun's distance—we find that he is hampered with a difficulty of precisely the same character.

The sun being an inaccessible object, the astronomer can apply no other methods to determine its distance—directly—than those which a surveyor would use in determining the distance of an inaccessible castle, or rock, or tree, or the like. We shall see presently that the ingenuity of astronomers has, in fact, suggested some other indirect methods. But clearly the most satisfactory estimate we can have of the sun's distance is one founded on such simple notions and involving in the main such processes of calculation as we have to deal with in ordinary surveying.

There is, in this respect, no mystery about the solution of the famous problem. Unfortunately, there is enormous difficulty.

When a surveyor has to determine the distance of an inaccessible object, he proceeds in the following manner: He first very carefully measures a base-line of convenient length. Then from either end of the base-line he takes the bearings of the inaccessible object—that is, he observes the direction in which it lies. It is clear that, if he were now to draw a figure on paper, laying down the base-line to some convenient scale, and drawing lines from its ends in directions corresponding to the bearings of the observed object, these lines would indicate, by their intersection, the true relative position of the object. In practice, the mathematician does not trust to so rough a method as construction, but applies processes of calculation.

Now, it is clear that in this plan every thing depends on the base-line. It must not be too short in comparison with the distance of the inaccessible object;

for then, if we make the least error in observing the bearings of the object, we get an important error in the resulting determination of the distances. The reader can easily convince himself of this by drawing an illustrative case or two on paper.

The astronomer has to take his base-line for determining the sun's distance, upon our earth, which is quite a tiny speck in comparison with the vast distance which separates us from the sun. It had been found difficult enough to determine the moon's distance with such a short base-line to work from. But the moon is only about a quarter of a million of miles from us, while the sun is more than ninety millions of miles off. Thus the problem was made several hundred times more difficult—or, to speak more correctly, it was rendered simply insoluble unless the astronomer could devise some mode of observing which should vastly enhance the power of his instruments.

For, let us consider an illustrative case. Suppose there were a steeple five miles off, and we had a base-line only two feet long. That would correspond as nearly as possible to the case the astronomer has to deal with. Now, what change of direction could be observed in the steeple by merely shifting the eye along a line of two feet? There is a ready way of answering. Invert the matter. Consider what a line of two feet long would look like if viewed from a distance of five miles. Would its length be appreciable, to say nothing of its being measurable? Yet it is just such a problem as the measurement of that line which the astronomer would have to solve.

But even this is not all. In our illustration only one observer is concerned, and he would be able to use one set of instruments. Suppose, however, that from one ènd of the two-feet line an observer using one set of instruments took the bearing of the steeple; and that, half a year after, another observer brought another set of instruments and took the bearing of the steeple from the other end of the two-feet line, is it not obvious how enormously the uncertainty of the result would be increased by such an arrangement as this? One observer would have his own peculiar powers of observation, his own peculiar weaknesses; the other would have different peculiarities. One set of instruments would be characterized by its own faults or merits, so would the other. One series of observations would be made in summer, with all the disturbing effects due to heat; the other would be made in winter, with all the disturbing effects due to cold.

The observation of the sun is characterized by all these difficulties. Limited to the base-lines he can measure on earth, the astronomer must set one observer in one hemisphere, another in the other. Each observer must have his own set of instruments; and every observation which one has made in summer will have to be compared with an observation which the other has made in winter.

Thus we can understand that astronomers should have failed totally when they attempted to determine the sun's distance without aid from the other celestial bodies.

It may seem at first sight as though nothing the other celestial bodies could tell the astronomer would be of the least use to him, since these bodies are for the most part farther off than the sun, and even those which approach nearest to us are still far beyond the limits of distance within which the simple plan followed by surveyors could be of any service. And besides, it might be supposed that information about the distance of one celestial body could be of no particular service toward the determination of the distance of another.

But two things aid the astronomer at this point: First of all, he has discovered the law which associates together the distances of all the planets from the sun; so that if he can determine the distance of any one planet he learns immediately the distances of all. Secondly, the planets in their motion travel occasionally into such positions that they become mighty indices, tracing out on a natural dial-plate the significant lesson from which the astronomer hopes to learn so much. To take an instance from the motions of another planet than the one we are dealing with. Mars comes sometimes so near the earth that the distance separating us from him is little more than one-third of that which separates us from the sun. Suppose that, at such a time, he is seen quite close to a fixed star. That star gives the astronomer powerful aid in determining the planet's distance. For, to observers in some parts of the earth, the planet will seem nearer to the star than he will to observers elsewhere. A careful comparison of the effects thus exhibited will give significant evidence respecting the distance of Mars. And we see that the star has served as a fixed mark upon the vast natural dial of the heavens, just as the division-marks on a clock-face serve to indicate the position of the hands.

Now, we can at once see why Venus holds so important a position in this sort of inquiry. Venus is our nearest neighbor among the planets. She comes several millions of miles nearer to us than Mars, our next neighbor on the other side. That is the primary reason of her being so much considered by astronomers. But there is another of equal importance. Venus travels nearer than our earth to the sun. And thus there are occasions when she gets directly between the earth and the sun. At those times she is seen upon his face, and his face serves as a dial-plate by which to measure her movements. When an observer at one part of the earth sees her on one part of the sun's face, another observer at some other part of the earth will see her on another, and the difference of position, if accurately measured, would at once indicate the sun's distance. As a matter of fact, other modes of reading off the indications of the great dial-plate have to be adopted. Before proceeding to consider those modes, however, we must deal with one or two facts about Venus's movements which largely affect the question at issue.

Let us first see what we gain by considering the distance of Venus rather than that of the sun.

At the time of a transit Venus is of course on a line between the earth and the sun, and she is at somewhat less than a third of the sun's distance from us. Thus whatever effect an observer's change of place would produce upon the sun would be more than trebled in the case of Venus. But it must not be forgotten that we are to judge the motions of Venus by means of the dial-plate formed by the solar disk, and that dial-plate is itself shifted as the observer shifts his place. Venus is shifted three times as much, it is true; but it is only the balance of change that our astronomer can recognize. That balance is, of course, rather more than twice as great as the sun's change of place.

So far, then, we have not gained much, since it has been already mentioned that the sun's change of place is not measurable by any process of observation astronomers can apply.

It is to the fact that we have the sun's disk whereby

to measure the change that we must chiefly trust; and even that would be insufficient were it not for the fact that Venus is not at rest, but travels athwart the great solar dial-plate. We are thus enabled to make a time measurement take the place of a measurement of space. If an observer in one place sees Venus cross the sun's face at a certain distance from the centre, while an observer at another place sees her follow a path slightly farther from the centre, the transit will clearly seem longer to the former observer than to the latter.

This artifice of exchanging a measurement of time for one of space—or *vice versa*—is a very common one among astronomers. It was Edmund Halley, the friend and pupil of Sir Isaac Newton, who suggested its application in the way above described. It will be noticed that what is required for the successful application of the method is that one set of observers should be as far to the north as possible, another as far to the south, so that the path of Venus may be shifted as much as possible. Clearly the northern observers will see her path shifted as much to the south as it can possibly be, while the southern observers will see the path shifted as far as possible toward the north.

One thing, however, is to be remembered. A transit lasts several hours, and our observers must be so placed that the sun will not set during these hours. This consideration sometimes involves a difficulty. For our earth does not supply observing room all over her surface, and the very region where observation would be most serviceable may be covered by a widely-extended ocean. Then, again, the observing parties are being rapidly swayed round by the rotating earth; and it is often difficult to fix on a spot which may not, through this cause, be shifted from a favorable position at the beginning of the transit to an unfavorable one at the end.

Without entering on all the points of difficulty involved by such considerations as these, we may simply indicate the fact that the astronomer has a problem of considerable complexity to solve in applying Halley's mode of observation to a transit of Venus.

It was long since pointed out by the French astronomer Delisle that the subject may be attacked another way—that, in fact, instead of noticing how much longer the transit lasts in some places than in others, the astronomer may inquire how much earlier it begins or ends in some places than in others.

Here is another artifice, extremely simple in principle, though not altogether so simple in its application. Our readers must bear with us while we briefly describe the qualities of this second method, because in reality the whole question of the transit and all the points which have to be attended to in the equipment and placing of the various observing parties depend on these preliminary matters. Without attending to them—or at least to such primary points as we shall select—it would be impossible to form a clear conception of the circumstances with which astronomers are about to deal. There is, however, no real difficulty about this part of the subject, and we shall only ask of the reader to give his attention to it for a very brief space of time.

Suppose the whole of that hemisphere of the earth on which the sun is shining when the transit is about to begin were covered with observers waiting for the event. As Venus sweeps rapidly onward to the critical part of her path, it is clear that some of these observers will get an earlier view of the commencement of the transit than others will; just as at a boat-race, persons variously placed round a projecting corner of the course see the leading boat come into view at different times. Some one observer on the outer rim of the hemisphere would be absolutely the first to see the transit begin. Then rapidly other observers would see the phenomenon; and in the course of a few minutes some one observer on the outer rim of the hemisphere—almost exactly opposite the first—would be absolutely the last to see the transit begin. From that time the transit would be seen by all for several hours—we neglect the earth's rotation, of course—but the end of the transit, like the beginning, would not be seen simultaneously by the observers. First one would see it, then in succession the rest, and last of all an observer almost exactly opposite the first.

Now, here we have had to consider four observers who occupy exceptional positions. There is (1) the observer who sees the transit begin earliest, (2) the one who sees it begin latest, (3) the one who sees it end earliest, and (4) the one who sees it end latest. Let us consider the first two only. Suppose these two observers afterward compared notes, and found out what was the exact difference of time between their respective observations. Is it not clear that the result would at once afford the means of determining the sun's distance? It would be the simplest of all possible astronomical problems to determine over what proportion of her orbit Venus passed in the interval of time which elapsed between these observations; and the observers would now have learned that that portion of Venus's orbit is so many miles long, for they know what distance separated them, and it would be easy to calculate how much less that portion of Venus's orbit is. Thus they would learn what the length of her whole orbit is, thence her distance from the sun, and thence the sun's distance from us.

The two observers who saw the transit end earliest and latest could do the like.

Speaking generally, and neglecting all the complexities which delight the soul of the astronomer, this is Delisle's method of utilizing a transit. It has obviously one serious disadvantage as compared with the other. An observer at one side of the earth has to bring his observations into comparison with those made by an observer at the other side of the earth. Each uses the local time of the place at which he observes, and it has been calculated that for the result to be of value there must not be an error of a single second in their estimates of local time. Now, does the reader appreciate the full force of this proviso? Each observer must know so certainly in what exact longitude he is, that his estimate of the time when true noon occurs shall not be one second wrong! This is all satisfactory enough in places where there are regular observatories. But matters are changed when we are dealing with such places as Woahoo, Kerguelen Land, Chatham Island, and the wilds of Siberia.

Here, however, as in so many other cases, the astronomer must take what he can get and be thankful. If Nature insists on not revealing her secrets unless astronomers will betake themselves to all manner of desert and uncanny places, all astronomers can do is to face with boldness the difficulties thus placed in their way, and to do their utmost to bring them into complete subjection.

In the coming transit there are many such difficulties to be encountered. In fact, it is almost impossible to conceive a transit the circumstances of which are more inconvenient. On the other hand, however, the transit is of such a nature that if once the preliminary difficulties are overcome, we can hope more from its indications than from those of any other transit which will happen in the course of the next few centuries.

The transit will begin earliest for observers in the neighborhood of the Sandwich Islands, latest for observers near Crozet Island, far to the southeast of the Cape of Good Hope. It ends earliest for observers far to the southwest of Cape Horn, latest for observers in the northeastern parts of European Russia. Thus we see that, so far as the application of our second method is concerned, the suitable spots are not situated in the most inviting regions of the earth's surface. As the transit happens on December 8, 1874, the principal northern stations will be very bleak abodes for the observers. The southern stations are in yet more dreary regions — notwithstanding the fact that the transit occurs during the summer of the southern hemisphere.

For the application of Halley's method we require stations where the whole transit will be visible, and, as the days are very short at the northern stations in December, it is as respects these that we encounter most difficulty. However, it has been found that many places in Northern China, Japan, Eastern Siberia, and Mantchooria are suitable for the purpose. The best southern stations for this method lie, unfortunately, on the unexplored Antarctic Continent and the islands adjacent to it; but Crozet Island, Kerguelen Land, and some other places more easy of access than the Antarctic Continent, will serve very well. Indeed, England has so many stations to occupy elsewhere that it is doubtful whether she will care to undertake the dangerous and difficult task of exploring the Antarctic wastes to secure the best southern stations. The work may fairly be left to other nations, and doubtless will be efficiently carried out.

What England will actually undertake has not yet been fully decided upon. We may be quite certain that she will send out a party to Woahoo or Hawaii to observe the accelerated commencement of the transit. She will also send observers to watch the retarded commencement, but whether to Crozet Island, Kerguelen Land, Mauritius, or Rodriguez, is uncertain. Possibly two parties will be sent out for this purpose, and most likely Crozet Island and Mauritius will be the places selected. It had been thought until lately that the sun would be too low at these places when the transit begins, but a more exact calculation of the circumstances of the transit has shown this to be a mistake. Both Crozet Island and Kerguelen Land are very likely to be enveloped in heavy mists when the transit begins—that is, soon after sunrise—hence the choice of Mauritius or Rodriguez as a secondary station.

England will also be called on to take an important part in observing the accelerated end of the transit.

A party will probably be sent to Chatham Island or Campbell Island, not far from New Zealand. It had been thought that at the former island the sun would be too low; but here, again, a more exact consideration of the circumstances of the transit has led astronomers to the conclusion that the sun will be quite high enough at this station.

The Russian observers are principally concerned with the observation of the retarded end of the transit, nearly all the best stations lying in Siberia. But there are several stations in British India where this phase can be very usefully observed; and doubtless the skilful astronomers and mathematicians who are taking part in the survey of India will be invited—as at the time of the great eclipse—to give their services in the cause of science. Alexandria, also, though inferior to several of the Indian stations, will probably be visited by an observing party from England.

It will be seen that England will thus be called on to supply about half a dozen expeditions to view the transit. All of these will be sent out in pursuance of Delisle's mode of utilizing a transit, so that, for reasons already referred to, it will be necessary that they should be provided with instruments of the utmost delicacy, and very carefully constructed.* They will

* It is held to be of the utmost importance that all the observing parties should use similar telescopes. It would be well if the class of telescope selected were Browning's six-inch reflectors.

have to remain at their several stations for a long time before the transit takes place—several months, at least—so that they may accurately determine the latitude of the temporary observatories they will erect. This is a work requiring skilled observers and recondite processes of calculation. Hence it is that the cost of sending out these observing parties is so considerable.

The only English party which will apply Halley's method of observation is the one which will be stationed at Crozet Island or Kerguelen Land. This part of their work will be comparatively easy, the method only requiring that the duration of the transit should be carefully timed. In fact, one of the great advantages of Halley's method is the smallness of the expense it involves. A party might land the day before the transit and sail away the day after, with results at least as trustworthy as those which a party applying Delisle's method could obtain after several months of hard work. It is to this, rather than any other cause, that the small expense of the observations made in 1769 is to be referred. And doubtless had it been decided by our astronomical authorities to apply Halley's method solely or principally, the expense of the transit-observations would have been materially lessened. There would, however, have been a risk of failure through the occurrence of bad weather at the critical stations; whereas now—as other nations will doubtless avail themselves of Halley's method—the chance that the transit-observations will fail through meteorological causes is very largely diminished. Science will owe much to the generosity of England in this respect.

It is, indeed, only recently that the possibility of applying Halley's method has been recognized. It had been thought that the method must fail totally in 1874. But on a more careful examination of the circumstances of the transit, a French astronomer, M. Puiseux, was enabled to announce that this is not the case. Almost simultaneously the present writer published calculations pointing to a similar result; but having carried the processes a few steps further than M. Puiseux, he was able to show that Halley's method is not only available in 1874, but is the more powerful method of the two.

Unfortunately, there is an element of doubt in the inquiry, of which no amount of care on the part of our observers and mathematicians will enable them to get rid. We refer to the behavior of Venus herself. It is to the peculiarity we are now to consider that the *quasi*-failure of the observations made in 1769 must be attributed. It is true that Mr. Stone, the eminent first-assistant at the Greenwich Observatory, has managed to remove the greater part of the doubts which clouded the results of those observations. But not even his skill and patience can serve to remove the blot which a century of doubt has seemed to throw upon the most exact of the sciences. We shall now show how much of the blame of that unfortunate century of doubt is to be ascribed to Venus.

At a transit, astronomers confine their attention to one particular phase—the moment, namely, when Venus just seems to lie wholly within the outline of the sun's disk. This at least was what Halley and Delisle both suggested as desirable. Unfortunately, Venus had not been consulted, and when the time of the transit came she declined to enter upon or leave the sun's face in the manner suggested by the astronomers. Consider, for example, her conduct when entering on the sun's face:

At first, as the black disk of the planet gradually notched the edge of the sun's disk, all seemed going on well. But when somewhat more than half of the planet was on the sun's face, it began to be noticed that Venus was losing her rotundity of figure. She became gradually more and more pear-shaped, until at last she looked very much like a peg-top touching with its point the edge of the sun's disk. Then suddenly—"as by a lightning-flash," said one observer—the top lost its peg, and then gradually Venus recovered her figure, and the transit proceeded without further change on her part until the time came for her to leave the sun's face, when similar peculiarities took place in a reversed order.

Here was a serious difficulty indeed. For when was the moment of true contact? Was it when the peg-top figure seemed just to touch the edge of the sun? This seemed unlikely, because a moment after the planet was seen well removed from the sun's edge. Was it when the rotund part of the planet belonged to a figure which would have touched the sun's edge if the rotundity had been perfect elsewhere? This, again, seemed unlikely, because at this moment the black band connecting Venus and the sun was quite wide. And, besides, if this were the true moment of contact, what eye could be trusted to determine the occurrence of a relation so peculiar? Yet the interval between this phase and the final or peg-top phase lasted several seconds—as many as twenty-two in one instance in 1769—and the whole success of the observation depended on exactness within three or four seconds at the outside.

We know that Venus will act in precisely the same manner in 1874. If we had been induced to hope that improvements in our telescopes would diminish the peculiarity, the observations of the transit of Mercury in November, 1868, would have sufficed to destroy that hope, for, even with the all but perfect instruments of the Greenwich Observatory, Mercury assumed the peg-top disguise in the most unpleasing manner.

It may be asked, then, What do astronomers propose to do in 1874 to prevent Venus from misleading them again as she did in 1769? Much has already been done toward this end. Mr. Stone undertook a series of careful researches to determine the law according to which Venus may be expected to behave or to misbehave herself; and the result is, that he has been able to tell the observers exactly what they will have to look for, and exactly what it is most important that they should record. In 1769, observers recorded their observations in such doubtful terms, owing to their ignorance of the real significance of the peculiarities they witnessed, that the mathematicians who had to make use of those observations were misled. *Hinc illæ lacrymæ.* Hence it is that an undeserved reproach has fallen upon the "exact science."

The amount of the error resulting from the misinterpretation of the observations made in 1769 was, however, very small indeed, when its true character is considered. It is, indeed, easy to make the error seem enormous. The sun's distance came out some four millions of miles too large, and that seems no trifling error. Then, again, the resulting estimate of the distance of Neptune came out more than a hundred million miles too great; while even this enormous error was as nothing when compared with that which resulted when the distances of the fixed stars were considered.

But this is an altogether erroneous mode of estimating the effect of the error. It would be as absurd to count up the number of hairs' breadth by which the geographer's estimates of the length and breadth of

England may be in error. In all such matters it is relative and not absolute error we have to consider. A microscopist would have made a bad mistake who should over-estimate the length of a fly's proboscis by a single hair's breadth; but the astronomer had made a wonderfully successful measurement of the sun's distance who deduced it within three or four millions of miles of the true value. For it is readily calculable that the error in the estimated relative bearing of the sun as seen from opposite sides of the earth corresponds to the angle which a hair's breadth subtends when seen from a distance of 125 feet.

The error was first detected when other modes of determining the sun's distance were applied by the skilful astronomers and physicists of our own day. We have no space to describe as fully as they deserve the ingenious processes by which the great problem has been attacked without aid from Venus. Indeed, we can but barely mention the principles on which those methods depend. But to the reader who takes interest in astronomy, we can recommend no subject as better worth studying than the masterly researches of Foucault, Leverrier, Stone, and Hansen, upon the problem of the sun's distance.

The problem has been attacked in four several ways. First, the tremendous velocity of light has been measured by an ingenious arrangement of revolving mirrors; the result combined with the known time occupied by light in travelling across the earth's orbit immediately gives the sun's distance. Secondly, a certain irregularity in the moon's motion, due to the fact that she is most disturbed by the sun when traversing that half of her path which is nearest to him, was pressed into the service with similar results. Thirdly, an irregularity in the earth's motion, due to the fact that she circles around the common centre of gravity of her own mass and the moon's, was made a means of attacking the problem. Lastly, Mars, a planet which, as we have already mentioned, approaches us almost as nearly as Venus, was found an efficient ally.

The result of calculations founded on these methods showed that the sun's distance, instead of being about 95,000,000 miles, is little more than 91,500,000 miles. And recently, by a careful reëxamination of the observations made upon Venus in 1769, Mr. Stone has shown that they point to a similar result.

Doubtless, however, we must wait for the transit of Venus in 1874 before forming a final decision as to the estimate of the sun's distance which is to take its place in popular works on astronomy during the next century or so. Nothing but an unlooked-for combination of unfavorable circumstances can cause the failure of our hopes. Certainly, if we should fail in obtaining satisfactory results in 1874, the world will not say that the generosity of the English Government has been in fault, since it would be difficult to find a parallel in the history of modern science to the munificence of the grant which has been made this year for expeditions to observe a phenomenon whose interest and importance are purely scientific.

(From *St. Paul's*, October, 1869.)

---

## *RECENT SOLAR RESEARCHES.*

SINCE the great eclipse of August, 1868, our knowledge respecting the constitution of the sun has been steadily progressing. One discovery after another has been made, and there really seems to be no reason for believing that we have as yet nearly reached the limits of the knowledge which spectroscopic analysis is capable of supplying. Indeed, the invention of a new form of spectroscope—the ingenious automatic spectroscope of Mr. Browning—promises soon to be rewarded by a series of discoveries as important as any which have hitherto been made. We propose briefly to indicate the present position of our knowledge respecting the great central luminary of our system.

The spectroscopic observation of the eclipse of August, 1868, had shown that the strange prominences seen during total eclipses of the sun are vast masses of luminous vapor—hydrogen-flames, we may call them, considering how largely hydrogen enters into their constitution. Only we must remember that it is hydrogen glowing from intensity of heat simply, and not burning hydrogen, that constitutes these prominences. Now, it had long been recognized that the colored prominences spring from an envelope of a similar nature surrounding the whole surface of the sun. Father Secchi, of the Collegio Romano, in a lecture given to the pupils of the École Ste. Geneviève, had thus in 1867 described this envelope (whose existence he was the first to recognize): "The observation of eclipses furnishes indisputable evidence that the sun is really surrounded by a layer of red matter, of which we commonly see no more than the most elevated points." One of the first and most interesting results of the eclipse-observations was Mr. Lockyer's confirmation of the justice of this opinion. He and Jannsen had independently shown that the existence of prominences can be recognized when the sun is not eclipsed; and the same method supplied clear evidence of the existence of this red envelope, to which Mr. Lockyer gave the name of the *chromosphere.* Remembering who first indicated its existence as "indisputable," we may conveniently call it Secchi's chromosphere. (See *note* at the end of this paper.)

Both the chromosphere and the prominences consist of glowing vapor. But there is a difference in their constitution. In the prominences there are usually but very few constituent vapors. Hydrogen is there, and another vapor, whose nature is as yet undetermined, while occasionally there are the vapors of other elements. But in the chromosphere there are commonly several elements, and sometimes there are many.

Here, then, we have above the photosphere of the sun a vaporous envelope, obviously of a complicated structure, and perhaps far more complicated than it has yet been proved to be. For it must be remembered that the lowest layers of this envelope might be composed of the vapors of numerous elements, and yet no record of their existence be recognized. A depth of ten miles would correspond to so small a portion of the sun's diameter (about the 85,000th part) as to be wholly unrecognizable by any telescopic power men can hope to obtain. If any of our readers are telescopists, they will know what force lies in the remark that such a distance would subtend about the 44th part of a second of arc, so that no less than twenty-six such distances could be placed between the components of that well-known test-object, the double companion of the star Gamma Andromedæ.*

Next below this colored envelope there is the mottled photosphere, either a white-hot surface with

* The view here presented was completely confirmed during the eclipse of last December. Professor Young and Mr. Pye independently recognized a layer whose spectrum showed all the Fraunhofer lines reversed. By observing at the place where the moon had just concealed the last fine sickle of the solar disk, they obviated the effects of diffraction, which render the observation wholly impossible in the case of the uneclipsed sun.

relatively dark pores all over it, or, according to other and better authorities, a surface of white-hot spots spread over a relatively dark background. Here we are describing merely its appearance; what the constitution of this surface may in reality be remains yet to be determined.

Beneath the photosphere there are vast depths of vapor, for when the photosphere is broken through where spots are formed, the spectroscope tells us that the relatively dark regions thus disclosed are filled with the vapors of various elements. We know that the dark lines which cross the rainbow-tinted solar spectrum are caused by the light-absorbing action of the vapors which surround the sun, and these lines are seen more distinctly in the spectrum of a sun-spot than in that of the photosphere.

Now, it is worthy of notice that all that has thus far been discovered tends to confirm the theroy put forward nearly a century ago by Sir William Herschel. That thoughtful observer recognized in the solar photosphere a widely-extended layer of luminous clouds, while he regarded the light of the penumbræ of sun-spots as coming from a lower cloud-layer. He conceived that up-rushes of vapor, thrusting aside both layers, caused the appearance of a solar spot. We have heard a great deal lately of the English and Continental theories of the solar constitution; but the evidence we have recently obtained goes far to show that, after all, Sir William Herschel, without the aid of spectroscope or polariscope, formed a juster view of the solar constitution than any which has been recently propounded. He was doubtless mistaken in the view (which he put forward as a mere hypothesis) that the real surface of the sun may be not very intensely heated. We have every reason to believe that the whole mass of the sun is raised to an inconceivable degree of heat. But for the rest, there seems far more reason to believe in Sir William Herschel's cloud-layer theory than in any other which has been put forward in recent times.

Let us consider some of the consequences of such a constitution. Imagine the ascent of vapors of many elements from the fluid surface of the solar oceans. This mixed atmosphere is in reality aglow with the intensest heat and light, so that, if we could examine its spectrum separately, we should see the bright lines of the various vaporous elements which constitute it. But intensely hot as it is, it must yet be less hot than the surface from which it has risen, because the formation of vapors is a process in which heat is used up. And therefore, by a well-known law, the spectrum of the light from the white-hot surface shining through the atmosphere will be a rainbow-tinted streak, crossed by the dark lines corresponding to the various elements composing that atmosphere. But as the lighter vapors in this mixed atmosphere ascend, they reach a region of less pressure, and a region where they can part more freely with their heat. Thus, precisely as the cumulus-clouds form in our own atmosphere, so would a layer of clouds be formed somewhat low down in the solar atmosphere. But from the upper surface of this layer the vapors of the elements composing the clouds would rise, again to condense at a higher level, much as the light cirrus-clouds in our own atmosphere form at a great height above the layer of cumulus-clouds.

The great difference between this process and what takes place in our own atmosphere would consist in the fact that whereas the only kind of cloud which can form in our air is a water-cloud, there can be formed in the solar atmosphere clouds of iron, copper, zinc, and other such elements, each element having its own distinct range, so to speak, within the limits of the solar atmosphere.

Now, with such processes as these going on, we can conceive how rushes of heated gas might from time to time thrust aside the cloud-layers; and how where this happened we should occasionally recognize the bright lines corresponding to the more intensely-heated gas, as well as the dark lines corresponding to the deep vapor-masses laid bare by the removal of the photosphere. And precisely in this way do the observations recently made by Mr. Lockyer seem alone to be explicable. He sees the glowing vapors above the photosphere stirred from time to time as by fierce tempests—

nay, he is enabled to measure (very roughly, of course) the velocity with which these solar winds urge their way through the chromosphere itself, in the neighborhood of these spots. The progress of these hurricanes is often indicated by the appearance of bright lines in those parts of the spectrum where usually dark lines are seen.

Truly Kirchhoff's discovery of the significance of the spectral lines is bearing wonderful fruit! Who would have thought that researches carried on with a few triangular prisms of glass on the light from such a substance as sodium, the basis of our commonplace soda, would lead to the result that solar tornadoes could be watched as readily with the spectroscope as in Galileo's time the sun-spots themselves could be traced across the sun's disk with the telescope? *

(From the *Spectator* for July 2, 1870.)

* I give this paper as it appeared in the *Spectator*. But there are some points requiring correction. In the first place, the objectionable word *chromosphere* (for *chromatosphere*) should be replaced by *sierra*. Secondly, there is an error as to the absolute priority of Secchi in recognizing the sierra. He went considerably beyond all others in the matter, having not only reasoned upon, but seen and photographed the sierra, and having furthermore found evidence as to its nature when studying sun-spots. But Professors Grant and Swan, as well as Von Littrow, the Imperial Astronomer of Austria, had recognized the existence of the sierra before Secchi, and Leverrier had also independently arrived at the same conclusion as Secchi, and at about the same time. I had not known of some of these claims and had forgotten others when I wrote the above paper. This will scarcely seem surprising when it is remembered that the views of Grant, Swan, Von Littrow, and Leverrier, had not been made widely public—as Secchi's had—by being published in popular treatises

## *GOVERNMENT AID TO SCIENCE.*

Among the questions which will occupy the attention of the new Parliament, we think we may safely include, in anticipation, the subject of State intervention to secure the progress of physical science.* It will be remembered that this subject was brought before the notice of the British Association, at its recent meeting, by Lieutenant-Colonel Strange, and a committee—including the names of Professors Sir William Thomson, Tyndall, Frankland, Williamson, Stokes, Fleming, Jenkins, Hirst, and Huxley, Messrs. Glaisher and Huggins, and Drs. Stenhouse, Balfour Stewart, and Mann—was appointed to consider and report upon the subject. Science has now reached a peculiar stage

and in lectures. It was with some surprise, therefore, that I found myself charged, not only with ignorance, but, incongruously enough, with injustice also, by a fellow-worker in astronomy, who addressed a letter to the editor of the *Spectator*, advocating in needlessly warm terms the prior claims of Grant and Swan. It is perhaps unnecessary for me to say that the charge of injustice was wholly undeserved; and I think the writer of the letter would have inferred this had he considerred a parallel instance which had recently occurred. For a well-known worker had claimed the very same discovery only a few months before as *his own;* and, although the subject was specially his, he had not known even of Secchi's numerous public statements respecting the sierra, yet no one thought of charging him with injustice. The writer of the letter could scarcely have forgotten the circumstance, since that worker was no other than himself.

* The reader need hardly be told that the hopes here expressed were completely disappointed.

in that long and remarkable career of progress which was inaugurated toward the close of the sixteenth century. Hitherto those who have been able and willing to take part in scientific researches have had the means of doing so without incurring great expense, and many have even found it possible to do good and useful service in the cause of science while prosecuting, at the same time, the labors of their profession or trade. But now the case is very different. A man who would assist in forwarding the progress of science must give his whole energies to the cause; he must be prepared to incur large expenses; and all this he must do without the hope that science will make him any pecuniary return. Theoretically, indeed, it may be argued that he will labor best who hopes for no return for his labors; who works, not for profit, but from pure love of science, and so on. But, as a matter of fact, many of those who would serve science best, and hundreds of those who could do yeoman's service in her cause, are simply debarred from scientific pursuits by the necessity of earning the means of subsistence. And there are crowds of others who, though they may be independent in means, are yet unable to provide themselves with the expensive instruments by which alone any useful work can now be done. For, as Colonel Strange observes, "Science can no longer be cultivated as in by-gone times it used to be. In astronomy the man with his table spy-glass cannot now furnish acceptable results. In chemistry, the Wollaston tea-tray and wine-glasses are superseded by well-equipped laboratories. In optics we see elaborate spectroscopes, not Newton's simple prism. In meteorology, and in every investigation of continuous phenomena, we are satisfied with nothing less than self-recording instruments. In electricity, in microscopy, and in other branches, our appliances are every day more and more amplified. The age of great discoveries made, and, above all, extensive series of facts accumulated with limited means, is passing away; and we are every day compelled to employ more perfect appliances and more systematic agencies in unravelling the secrets of Nature."

It is scarcely necessary to point out that the aid of the State in securing the progress of physical science is not asked without the promise of a *quid pro quo*. It is not as though the State were called upon to aid in antiquarian, or entomological, or numismatic researches, or in any subject of inquiry which, however interesting, has no practical bearing on the wants of daily life, or on the appliances by which the social state of man may be benefited and improved. Nor is it to secure the spread of scientific knowledge that State aid is called for, but to secure the progress of physical science. That that progress cannot fail to bring with it important advantages to mankind it is almost needless to assert. We have only to look around us to see what science has done for mankind. But those are best acquainted with the practical value of scientific knowledge who are themselves engaged in scientific researches, or are at least proficient in scientific matters. Hundreds, for example, might see in the complicated instruments which are to be found in the Greenwich Observatory nothing but ingenious applications for the solution of theoretical problems; it is only astronomers, or those who are versed in the processes of astronomy, who know that our whole system of commerce would be affected injuriously if those instruments were destroyed or left unused. Here we have an instance of science working under State patronage, working in the cause of the State; and what Colonel Strange proposes is to multiply instances of this sort. The State profits by the labors of the Greenwich astronomers, and those astronomers would for the most part be unable or unwilling to continue their labors but for the pecuniary reward which they receive from the State. But assuredly the State would suffer more than the astronomers if the establishment at Greenwich were done away with. And precisely in the same way the State would reap important advantages from the labors of proficients in other departments of science who are now debarred by considerations of expense, or by the necessity of earning a livelihood, from applying their skill to forward the cause of scientific progress.

Colonel Strange's proposal includes the establishment of national institutions expressly for the practical advancement of scientific research. He remarks that "men engaged in science need hardly be told that when they discover a new substance, the determination of the physical properties of which is attended with cost and labor, they experience a great—perhaps insuperable—difficulty in obtaining its examination. A new theory, or the confutation or confirmation of an old one, if dependent on any considerable accumulation of facts, shares even a worse fate." Important benefits could not fail to result if difficulties such as these were removed from the paths of physical research, by the institution of bodies whose duty it would be to undertake, and complete in an accurate and systematic manner, costly and tedious investigations on which vast interests may be dependent.

(From the *Daily News* for December 9, 1869.)

---

## *AMERICAN ALMS FOR BRITISH SCIENCE.**

Our astronomers have received an invitation which is as pleasing to them as men of science as it is painful

* This was one of a series of articles which appeared in the *Daily News* during the months which followed the announcement that the British Government would give no aid to the eclipse expedition. To the liberality with which the *Daily News* gave space for these appeals may fairly be ascribed the fact that eventually the eclipse committee was aroused to something like energetic action. When the real state of the case became known to Government, ample assistance was rendered. The

to them as Englishmen. As our readers know, sixty-eight persons had volunteered to go to Spain and Sicily to view the total eclipse of December 22d; our scientific societies had voted large sums of money for the equipment of the two observing parties; and every one was certain that Government would supply the means of transport. But every one was mistaken. The Admiralty discovered that the nation would assuredly disapprove if room were found for mere men of science and their trumpery in any of her Majesty's ships; and accordingly, just when the extensive preparations requisite for the expeditions were in full progress, news came that the means of transport must be found by the observers themselves. We do not care here—we hardly have patience, indeed—to discuss the probable cause of a refusal so discreditable to the scientific repute of England. It had been announced by the Astronomer-Royal (in connection with another matter), that Government would always be found liberal in scientific matters, if a sufficient cause were shown by persons in whom they had trust; and we do not care to inquire whether the Astronomer-Royal was mistaken in this matter, or whether the Government declined to

shortness of the time eventually left for preparation may be regarded as accounting for subsequent seeming short-comings on the part of the Organizing Committee; while fortunately the zeal of the expeditionists averted the risk (which at one time seemed serious) that rather brusque usage would cause some of the most important members of the expeditions to withdraw their aid.

put trust in him or in the Presidents of our Astronomical and Royal Societies, or whether, lastly, the sufficient cause was not brought before the Government with proper earnestness. Let the explanation be what it may, the fact remains—England has been exhibited to all the nations as turning her back on science, and English men of science have been discredited before the world as unworthy of England's confidence.

But now news comes that the Government of the United States has not only found means of transport for two American parties, but has made the handsome grant of £6,000, to furnish suitable appliances for observing the eclipse. The American men of science have reached England. They recognize the pitiable condition to which our astronomers have been reduced by the Government, and they invite our sixty-eight volunteers to sail with them. A letter has been sent to these volunteers, inviting them, in the name of the American expeditionary parties, to accept this much-needed assistance. The offer is most generous; it is most inviting; it is one which no astronomer is justified in declining on account of sentimental considerations. But it certainly is a new and a painful position for an English man of science to be placed in, thus to find scientific alms offered him as a reparation for the insult he has, in effect, received from his own Government.

Many may be disposed to wonder why so much interest is attached to this particular eclipse. During many former total eclipses—even when they have been visible at more conveniently accessible stations—less care was taken to fit out expeditions. And, what is even more to the point, observations have been made on eclipse after eclipse, in former times, without adding jot or tittle to our knowledge of solar physics. But during recent eclipses things have altered. In 1860 the celebrated "Himalaya Expedition" sailed to Spain from England; while other parties came from France, Italy, and Germany. And, though the old fault of wasting observing energy on matters already known or demonstrated prevailed very largely, yet De la Rue and Secchi, by photographing the eclipsed sun, well repaid the whole cost of these expeditions. In the great total eclipse of August, 1868, Europe sent out many observing parties to India, and the great discovery that the red prominences seen round the totally-eclipsed sun are masses of glowing vapor sufficiently repaid the cost. In August, 1869, the Americans availed themselves right skilfully and worthily of the passage of the moon's shadow across their continent; and, though they failed in the main purpose they had set themselves, they made preliminary observations of the utmost importance and value. That purpose was to ascertain the nature of the glorious aureole of light seen around the sun during total eclipses; and it is with the same purpose that the expeditions formed for observing the present eclipse were to have set forth. The questions to be answered are full of interest, even now when their full significance is not known; while it may well be that when we begin to have accurate information about them, we shall find they have a real importance wholly unlooked for. As the last direct rays of the sun are concealed by the advancing moon, there springs into view a glorious crown of colored light—pearly white in parts, faintly pink beyond, and at the extreme verge showing tints of mauve and violet and green—delicate and beautiful beyond description. Through this coronal glory there extend rays of bluish-white light, reaching often to a vast distance from the black disk of the moon. Commonly remaining unchanged in position, these rays sometimes—if all the narratives can be trusted—exhibit very obvious signs of motion, resembling in this respect those streamers of colored light which we have lately so much admired in the aurora. Indeed, wonderful as it may seem, the corona has lately come to be regarded as associated in some way with the Aurora Borealis. We know that those auroral streamers which form so wonderful a display in our own skies are due to solar influences. In whatever way it may be brought about, certain it is that disturbances of the sun are reflected in terrestrial auroral displays. The auroras which have occurred lately were predicted by astronomers, who know that the sun is undergoing during the present year disturbances of the most amazing nature. Solar spots, of various dimensions, have been counted by the hundred of late; and we now know that when the sun is thus spotted our earth sympathizes with the central orb. Thrilling from pole to pole in magnetic tremor, she spreads out over both hemispheres the auroral banners that indicate the progress of electric revolutions. The devices of her children for utilizing her electric forces are for the time set at naught, and the telegraph-clerk finds for a while that Mother Earth is having her own way and will not obey his behests. If the sun, ninety millions of miles away from us, thus affects the earth's frame, and thus illuminates terrestrial skies, it need not be greatly wondered at should it be proved that he illuminates with no dissimilar light the regions lying more closely around him. If there are no planets like our earth in these regions, no large bodies on which the sun can exert his inconceivable powers, there are yet in these spaces —unless astronomers are at fault—uncounted millions of minute bodies, those tiny "pocket-planets" which pass at times through our own atmosphere, and are called by us falling stars, or meteors. Among these tiny bodies auroral gleams may pass, producing by their united lustre the glories of the solar corona.

But, whether this view be just, or whether, as Mr. Lockyer holds, the corona is only a phenomenon of our own air, or is due (as the fanciful M. Faye once thought of the colored prominences) to some sort of lunar mirage, certain it is that just now it is a matter of extreme interest that further observations should be made. Undoubtedly, what we have lately learned respecting the sun gives an interest and importance to this matter of the solar corona which it never before possessed. Yet this is the problem respecting which our Government is understood to have said to astronomers, "As far as we can, we will prevent you from solving it."

Truly it would be difficult to show that any material profit can be gained by solving the problems associated with the solar corona. The tree of science has its blossoms as well as its fruits, and perhaps the results of the observations we are advocating will belong to the former rather than the latter. But what then? Can we limit science to remunerative researches alone? As well might we attempt to get fruit from a tree whose leaves and blossoms we systematically plucked off. Latent though the power of science now is in great part, yet science is the greatest power our country possesses. It has been treated for a long while as a troublesome beggar—a few hundreds doled out here and a few thousands there. The country does not yet know its own interest. Because little has been asked, it has thought little could be returned. The time is coming when not hundreds or thousands will be asked for science, but millions freely and eagerly given—when the example of other countries, rapidly passing in advance of England through their scientific resources, will force on our attention the folly of a system which grants thirty millions yearly to secure the means of carrying on war, and refuses a few paltry thousands to secure the noblest portion of our strength.

(From the *Daily News* for November 5, 1870.)

---

## *THE SECRET OF THE NORTH POLE.*

If an astronomer upon some distant planet has ever thought the tiny orb we inhabit worthy of telescopic study, there can be little doubt that the snowy regions which surround the arctic and antarctic poles must have attracted a large share of his attention. Waxing and waning with the passing seasons, those two white patches afford significant intelligence respecting the circumstances of our planet's constitution. They mark the direction of the imaginary axial line upon which the planet rotates; so that we can imagine how an astronomer on Mars or Venus would judge from their position how it fares with terrestrial creatures. There may, indeed, be Martial Whewells who laugh to scorn the notion that a globe so inconveniently circumstanced as ours can be inhabited, and are ready to show that if there were living beings here they must be quickly destroyed by excessive heat. On the other hand, there are doubtless skeptics on Venus also who smile at the vanity of those who can conceive a frozen world, such as this our outer planet, to be inhabited by any sort of living creature. But we doubt not that the more advanced thinkers both in Mars and Venus are ready to admit that, though we must necessarily be far inferior beings to themselves, we yet manage to "live and move and have our being" on this ill-conditioned globe of ours. And these, observing the earth's polar snow-caps, must be led to several important conclusions respecting physical relations here.

It is, indeed, rather a singular fact to contemplate, that ex-terrestrial observers, such as these, may know much more than we ourselves do respecting those mysterious regions which lie close around the two poles. Their eyes may have rested on spots which with all our endeavors we have hitherto failed to reach. Whether, as some have thought, the arctic pole is in summer surrounded by a wide and tide-swayed ocean; whether there lies around the antarctic pole a wide continent bespread with volcanic mountains larger and more energetic than the two burning cones which Ross found on the outskirts of this desolate region; or whether the habitudes prevailing near either pole are wholly different from those suggested by geographers and voyagers —such questions as these might possibly be resolved at once, could our astronomers take their stand on some neighboring planet, and direct the searching power of their telescopes upon this terrestrial orb. For this is one of those cases referred to by Humboldt, when he said that there are circumstances under which man is able to learn more respecting objects millions of miles away from him than respecting the very globe which he inhabits.

If we take a terrestrial globe, and examine the actual region near the North Pole which has as yet remained unvisited by man, it will be found to be far smaller than most people are in the habit of imagining. In nearly all maps the requirements of charting result in a considerable exaggeration of the polar regions. This is the case in the ordinary "maps of the two hemispheres" which are to be found in all atlases. And it is, of course, the case to a much more remarkable extent in what is termed Mercator's projection. In a Mercator's chart we see Greenland, for example, exaggerated into a continent fully as large as South America, or to seven or eight times its real dimensions.

There are three principal directions in which explorers have attempted to approach the North Pole. The first is that by way of the sea which lies between Greenland and Spitzbergen. We include under this head Sir Edward Parry's attempt to reach the pole by crossing the ice-fields which lie to the north of Spitzbergen. The second is that by way of the straits which lie to the west of Greenland. The third is that pursued by Russian explorers who have attempted to cross the frozen seas which surround the northern shores of Siberia.

In considering the limits of the unknown north-polar regions, we shall also have to take into account the voyages which have been made around the northern shores of the American Continent in the search for a "northwestern passage." The explorers who set out upon this search found themselves gradually forced to seek higher and higher latitudes if they would find a way round the complicated barriers presented by the ice-bound straits and islands which lie to the north of the American Continent. And it may be noticed in passing, as a remarkable and unforeseen circumstance, that the farther north the voyagers went the less severe was the cold they had to encounter. We shall see that this circumstance has an important bearing on the considerations we shall presently have to deal with.

One other circumstance respecting the search for the northwest passage, though not connected very closely with our subject, is so singular and so little known that we feel tempted to make mention of it at this point. The notion with which the seekers after a northwest passage set out was simply this, that the easiest way of reaching China and the East Indies was to pursue a course resembling as nearly as possible that on which Columbus had set out—if only it should appear that no impassable barriers rendered such a course impracticable. They quickly found that the American Continents present an unbroken line of land from high northern latitudes far away toward the antarctic seas. But it is a circumstance worth noticing, that if the American Continents had no existence, the direct westerly course pursued by Columbus was not only not the nearest way to the East-Indian Archipelago, but was one of the longest routes which could possibly have been selected. Surprising as it may seem at first sight, a voyager from Spain for China and the East Indies ought, if he sought the absolutely shortest path, to set out on an almost direct northerly route! He would pass close by Ireland and Iceland, and so, near the North Pole, and onward into the Pacific. This is what is called the great-circle route; and if it were only a practicable one, would shorten the course to China by many hundreds of miles.

Let us return, however, to the consideration of the information which arctic voyagers have brought us concerning the north-polar regions.

The most laborious researches in arctic seas are those which have been carried out by the searchers after a northwest passage. We will therefore first consider the limits of the unknown region in this direction. Afterward we can examine the results of those voyages which have been undertaken with the express purpose of reaching the North Pole along the three principal routes already mentioned.

If we examine a map of North America constructed in recent times, we shall find that between Greenland and Canada an immense extent of coast-line has been charted. A vast archipelago covers this part of the northern world. Or, if the strangely-complicated coast-lines which have been laid down really belong to but a small number of islands, the figures of these must be of the most fantastic kind. Toward the northwest, however, we find several islands whose outlines have been entirely ascertained. Thus we have in succession North Devon Island, Cornwallis Island, Melville Island, and Port Patrick Island, all lying north of the seventy-fifth parallel of latitude. But we are not to suppose that these islands limit the extent of our seamen's researches in this direction. Far to the northward of Wellington Channel, Captain de Haven saw, in 1852, the signs of an open sea—in other words, he saw, beyond the ice-fields, what arctic seamen call a "water-sky." In 1855 Captain Penny sailed upon this open sea; but how far it extends toward the North Pole has not yet been ascertained.

It must not be forgotten that the northwest passage has been shown to be a reality, by means of voyages from the Pacific as well as from the Atlantic. No arctic voyager has yet succeeded in passing from one ocean to the other. Nor is it likely now that any voyager will pursue his way along a path so beset by dangers as that which is called the northwest passage. Long before the problem had been solved, it had become well known that no profit could be expected to accrue to trade from the discovery of a passage along the perilous straits and the ice-encumbered seas which lie to the north of the American Continent. But Sir Edward Parry having traced out a passage as far as Melville Island, it seemed to the bold spirit of our arctic explorers that it might be possible, by sailing through Behring's Straits, to trace out a connection between the arctic seas on that side and the regions reached by Parry. Accordingly, McClure, in 1850, sailed in the "Investigator," and passing eastward, after traversing Behring's Straits, reached Baring's Land, and eventually identified this land as a portion of Banks's Land, seen by Parry to the southward of Melville Island.

It will thus be seen that the unexplored parts of the arctic regions are limited in this direction by sufficiently high latitudes.

Turn we next to the explorations which Russian voyagers have made to the northward of Siberia. It must be noticed, in the first place, that the coast of Siberia runs much farther northward than that of the American Continent. So that on this side, independently of sea explorations, the unknown arctic regions are limited within very high latitudes. But attempts have been made to push much farther north from these shores. In every case however, the voyagers have found that the ice-fields, over which they hoped to make their way, have become gradually less and less firm, until at length no doubt could remain that there lay an open sea beyond them. How far that sea may extend is a part of the secret of the North Pole; but we may assume that it is no narrow sea, since otherwise there can be little doubt that the ice-fields which surround the shores of Northern Siberia would extend unbroken to the farther shores of what we should thus have to recognize as a strait. The thinning-off of these ice-fields, observed by Baron Wrangel and his companions, affords, indeed, most remarkable and significant testimony respecting the nature of the sea which lies beyond. This we shall presently have to exhibit more at length; in the mean time we need only remark that scarcely any doubt can exist that the sea thus discovered extends northward to at least the eightieth parallel of latitude.

We may say, then, that from Wellington Channel, northward of the American Continent, right round toward the west, up to the neighborhood of Spitzbergen, very little doubt exists as to the general characteristics of arctic regions, save only as respects those unexplored parts which lie within ten or twelve degrees of the North Pole. The reader will see presently why we are so careful to exhibit the limited extent of the unexplored arctic regions in this direction. The guess we shall form as to the true nature of the north-polar secret will depend almost entirely on this consideration.

We turn now to those two paths along which arctic exploration, properly so termed, has been most successfully pursued.

It is chiefly to the expeditions of Drs. Kane and Hayes that we owe the important knowledge we have respecting the northerly portions of the straits which lie to the west of Greenland. Each of these explorers succeeded in reaching the shores of an open sea lying to the northeast of Kennedy Channel, the extreme northerly limit of those straits. Hayes, who had accompanied Kane in the voyage of 1854–'55, succeeded in reaching a somewhat higher latitude in sledges drawn by Esquimaux dogs. But both expeditions agree in showing that the shores of Greenland trend off suddenly toward the east at a point within some nine degrees of the North Pole. On the other hand, the prolongation of the opposite shore of Kennedy Channel was found to extend northward as far as the eye could reach. Within the angle thus formed there was an open sea "rolling," says Captain Maury, "with the swell of a boundless ocean."

But a circumstance was noticed respecting this sea which was very significant. The tides ebbed and flowed in it. Only one fact we know of—a fact to be presently discussed—throws so much light on the question we are considering as this circumstance does. Let us consider a little whence these tidal waves can have come.

The narrow straits between Greenland on the one side, and Ellesmere. Land and Grinnell Land on the other, are completely ice-bound. We cannot suppose that the tidal wave could have found its way beneath such a barrier as this. "I apprehend," says Captain Maury, "that the tidal wave from the Atlantic can no more pass under this icy barrier, to be propagated in the seas beyond, than the vibrations of a musical string can pass with its notes a fret on which the musician has placed his finger."

Are we to suppose, then, that the tidal waves were formed in the very sea in which they were seen by Kane and Hayes? This is Captain Maury's opinion: "These tides," says he, "must have been born in that cold sea, having their cradle about the North Pole."

But if we carefully consider the theory of the tides this opinion seems inadmissible. Every consideration on which that theory is founded is opposed to the assumption that the moon could by any possibility raise tides in an arctic basin of limited extent. It would be out of place to examine at length the principle on which the formation of tides depends. It will be sufficient for our purposes to remark that it is not to the mere strength of the moon's "pull" upon the waters of any ocean that the tidal wave owes its origin, but to the difference of the forces by which the various parts of that ocean are attracted. The whole of an ocean cannot be raised at once by the moon; but if one part is attracted more than another, a wave is formed. That this may happen, the ocean must be one of wide extent. In the vast seas which surround the Southern Pole there is room for an immensely powerful "drag," so to speak; for always there will be one part of these seas much nearer to the moon than the rest, and so there will be an appreciable difference of pull upon that part.

The reader will now see why we have been so careful to ascertain the limits of the supposed north-polar ocean, in which, according to Captain Maury, tidal waves are generated. To accord with his views, this ocean must be surrounded on all sides by impassable barriers either of land or ice. These barriers, then, must lie to the northward of the regions yet explored, for there is open sea communicating with the Pacific all round the north of Asia and America. It only requires a moment's inspection of a terrestrial globe to see how small a space is thus left for Captain Maury's land-locked ocean. We have purposely left out of consideration, as yet, the advances made by arctic voyagers in the direction of the sea which lies between Greenland and Spitzbergen. We shall presently see that on this side the imaginary land-locked ocean must be more limited than toward the shores of Asia or America. As it is, however, it remains clear that, if there were any ocean communicating with the spot reached by Dr. Kane, but separated from all communication—by open water—either with the Atlantic or with the Pacific, that ocean would be so limited in extent that the moon's attraction could exert no more effective influence upon its waters than upon the waters of the Mediterranean—where, as we know, no tides are generated. This, then, would be a tideless ocean, and we must look elsewhere for an explanation of the tidal waves seen by Dr. Kane.

We thus seem to have *prima facie* evidence that the sea reached by Kane communicates either with the Pacific or with the Atlantic, or—which is the most probable view—with both those oceans. When we consider the voyages which have been made toward the North Pole along the northerly prolongation of the Atlantic Ocean, we find very strong evidence in favor of the view that there is open-water communication in this direction, not only with the spot reached by Kane, but with a region very much nearer to the North Pole.

So far back as 1607, Hudson had penetrated within 8½° (or about 600 miles) of the North Pole on this route. When we consider the clumsy build and the poor sailing qualities of the ships of Hudson's day, we cannot but feel that so successful a journey marks this route as one of the most promising ever tried. Hudson was not turned back by impassable barriers of land or ice, but by the serious dangers to which the floating masses of ice and the gradually-thickening ice-fields exposed his weak and ill-manned vessel. Since his time, others have sailed upon the same track, and hitherto with no better success. It has been reserved to the Swedish expedition of last year to gain the highest latitudes ever reached in a ship in this direction. The steamship " Sofia," in which this successful voyage was made, was strongly built of Swedish iron, and originally intended for winter voyages in the Baltic. Owing to a number of delays, it was not until September 16th that the " Sofia " reached the most northerly part of her journey. This was a point some fifteen miles nearer the North Pole than Hudson had reached. To the north there still lay broken ice, but packed so thickly that not even a boat could pass through it. So late in the season, it would have been unsafe to wait for a change of weather and a consequent breaking up of the ice. Already the temperature had sunk 16° below the freezing-point; and the enterprising voyagers had no choice but to return. They made, indeed, another push for the north a fortnight later, but only to meet with a fresh repulse. An ice-block with which they came into collision opened a large leak in the vessel's side; and when after great exertions they reached the land, the water already stood two feet over the cabin-floor. In the course of these attempts, the depths of the Atlantic were sounded, and two interesting facts were revealed. The first was that the island of Spitzbergen is connected with Scandinavia by a submarine bank; the second was the circumstance that to the north and west of Spitzbergen the Atlantic is more than two miles deep!

We come now to the most conclusive evidence yet afforded of the extension of the Atlantic Ocean toward the immediate neighborhood of the North Pole. Singularly enough, this evidence is associated not with a sea-voyage, nor with a voyage across ice to the borders of some northern sea, but with a journey during which the voyagers were throughout surrounded as far as the eye could reach by apparently fixed ice-fields.

In 1827 Sir Edward Parry was commissioned by the English Government to attempt to reach the North Pole. A large reward was promised in case he succeeded, or even if he could get within five degrees of the North Pole. The plan which he adopted seemed promising. Starting from a port in Spitzbergen, he proposed to travel as far northward as possible in sea-boats, and then, landing upon the ice, to prosecute his voyage by means of sledges. Few narratives of arctic travel are more interesting than that which Parry has left of this famous "boat-and-sledge" expedition. The voyagers were terribly harassed by the difficulties of the way; and, after a time, that most trying of all arctic experiences, the bitterly cold wind which comes from out the dreadful north, was added to their trials. Yet still they plodded steadily onward, tracking their way over hundreds of miles of ice with the confident expectation of at least attaining to the eighty-fifth parallel, if not to the Pole itself.

But a most grievous disappointment was in store for them. Parry began to notice that the astronomical observation by which in favorable weather he estimated the amount of their northerly progress, showed a want of correspondence with the actual rate at which they were travelling. At first he could hardly believe that there was not some mistake; but at length the unpleasing conviction was forced upon him that the whole ice-field over which he and his companions had been toiling so painfully was setting steadily southward before the wind. Each day the extent of this set became greater and greater, until at length they were actually carried as fast toward the south as they could travel northward.

Parry deemed it useless to continue the struggle. There were certainly two chances in his favor. It was possible that the north wind might cease to blow, and it was also possible that the limit of the ice might soon be reached, and that his boats might travel easily northward upon the open sea beyond. But he had to consider the exhausted state of his men, and the great additional danger to which they were subjected by the movable nature of the ice-fields. If the ice should break up, or if heavy and long-continued southerly winds should blow, they might have found it very difficult to regain their port of refuge in Spitzbergen before winter set in or their stores were exhausted. Besides, there were no signs of water in the direction they had been taking. The water-sky of arctic regions can be recognized by the experienced seaman long before the open sea itself is visible. On every side, however, there were the signs of widely-extended ice-fields. It seemed, therefore, hopeless to persevere, and Parry decided on returning with all possible speed to the haven of refuge prepared for the party in Spitzbergen. He had succeeded in reaching the highest northern latitudes ever yet attained by man.

The most remarkable feature of this expedition, however, is not the high latitude which the party attained, but the strange circumstance which led to their discomfiture. What opinion are we to form of an ocean at once wide and deep enough to float an ice-field which must have been thirty or forty thousand square miles in extent? Parry had travelled upward of three hundred miles across the field, and we may fairly suppose that he might have travelled forty or fifty miles farther without reaching open water; also that the field extended fully fifty miles on each side of Parry's northerly track. That the whole of so enormous a field should have floated freely before the arctic winds is indeed an astonishing circumstance. On every side of this floating ice-island there must have been seas comparatively free from ice; and could a stout ship have forced its way through these seas, the latitudes to which it could have reached would have been far higher than those to which Parry's party was able to attain. For a moment's consideration will show that the part of the great ice-field where Parry was compelled to turn back must have been floating in far higher latitudes when he first set out. He reckoned that he had lost more than a hundred miles through the southerly motion of the ice-field, and by this amount, of course, the point he reached had been nearer the Pole. It is not assuming too much to say that a ship which could have forced its way round the great floating ice-field would certainly have been able to get within four degrees of the Pole. It seems to us highly probable that she would even have been able to sail upon open water to and beyond the Pole itself.

And when we remember the direction in which Dr. Kane saw an open sea—namely, toward the very region where Parry's ice-ship had floated a quarter of a century before—it seems reasonable to conclude that there is open-water communication between the seas which lie to the north of Spitzbergen and those which lave the northwestern shores of Greenland. If this be so, we at once obtain an explanation of the tidal waves which Kane watched day after day in 1855. These had no doubt swept along the valley of the Atlantic, and thence around the northern coast of

Greenland. It follows that, densely as the ice may be packed at times in the seas by which Hudson, Scoresby, and other captains, have attempted to reach the North Pole, the frozen masses must in reality be floating freely, and there must therefore exist channels through which an adventurous seaman might manage to penetrate the dangerous barriers surrounding the polar ocean.

In such an expedition, chance unfortunately plays a large part. Whalers tell us that there is great uncertainty as to the winds which may blow during an arctic summer. The icebergs may be crowded by easterly winds upon the shores of Greenland, or by westerly winds upon the shores of Spitzbergen, or, lastly, the central passage may be the most encumbered, through the effects of winds blowing now from the east and now from the west. Thus the arctic voyager has not merely to take his chance as to the route along which he shall adventure northward, but often, after forcing his way successfully for a considerable distance, he finds the ice-fields suddenly closing in upon him on every side, and threatening to crush his ship into fragments. The irresistible power with which, under such circumstances, the masses of ice bear down upon the stoutest ship has been evidenced again and again; though, fortunately, it not unfrequently happens that some irregularity along one side or the other of the closing channel serves as a sort of natural dock, within which the vessel may remain in comparative safety until a change of wind sets her free. Instances have been known in which a ship has had so narrow an escape in this way, and has been subjected to such an enormous pressure, that when the channel has opened out again, the impress of the ship's side has been seen distinctly marked upon the massive blocks of ice which have pressed against her.

Notwithstanding the dangers and difficulties of the attempt, and the circumstance that no material gains can reward the explorer, it seems not unlikely that before many months are passed the North Pole will have been reached. Last year two bold attempts were made—one by the Swedes, as already mentioned, the other by German men of science. In each case the result was so far successful as to give good promise for future attempts. This year both these nations will renew their attack upon the interesting problem. The German expedition will consist of two vessels, the "Germania" and the "Greenland." The former is a screw-steamer, of 126 tons, and well adapted to encounter the buffets of the ice-masses which are borne upon the arctic seas. The other is a sailing-yacht of 80 tons, and is intended to act as a transport-ship, by means of which communication may be kept up with Europe. The "Germania" will probably winter in high northern latitudes; and we should not be much surprised if before her return she should have been carried to the very Pole. Nor can the prospects of the Swedish expedition be considered less promising, when we remember that last year, though hampered by the lateness of the season and other difficulties, they succeeded in approaching the Pole within a distance only a few miles greater than that which separated Parry from the Pole in 1829.

Certainly England has reason to fear that before the year 1870 has closed she will no longer be able to claim that her flag has approached both Poles more nearly than the flag of any other nation. There are considerations which make the recent supineness of our country in the matter of arctic travel much to be regretted. In the winter of 1874 there will occur one of those interesting phenomena by which Nature occasionally teaches men useful lessons respecting her economy. We refer to the transit of Venus on December 8th in that year. One of the most effective modes of observing this transit will require that a party of scientific men should penetrate far within the recesses of the desolate Antarctic Circle. Where are the trained arctic seamen to be found who will venture upon this service? Most of our noted arctic voyagers have earned their rest; and, as Commander Davis said at a recent meeting of the Geographical Society, those who go for the first time into the arctic or antarctic solitudes are too much tried by the effects of the new experience to be fit to undertake important scientific labors. He spoke with special reference to the transit of 1882, for the observation of which there is (I have lately shown) small occasion to employ arctic voyagers. It is just possible that for the transit of 1874 trained explorers belonging to the old school of arctic travel may still be found. But if not, no time should be lost in supplying the deficiency. I have shown within the last few months that journeys to the antarctic regions will be required for this transit, and not for the later transit (as had been supposed). The Astronomer-Royal has expressed his desire that the discovery may be rendered available by suitable expeditions. "Every series of observations," he remarks, "which can really be brought to bear upon this important determination will be valuable." Therefore, for this reason alone, and even if the reputation of England in the matter of arctic travel were altogether worthless, it would be well that efforts should quickly be made to prepare crews and commanders for the work of 1874, by "sending them to school," as Commander Davis expressed it, "in the arctic seas."

(From *St. Paul's*, June, 1869.)

---

## *IS THE GULF STREAM A MYTH?*

The Gulf Stream has recently attracted a large share of the attention of our men of science. The abnormal character of the weather which we experienced last winter has had something to do with this. The influence of the Gulf Stream upon our climate, and the special influence which it is assumed to exercise in mitigating the severity of our winters, have been so long recognized that meteorologists began to inquire what changes could be supposed to have taken place in the great current to account for so remarkable a winter as the last. But it happened also that at a meeting of the Royal Geographical Society early in the present year the very existence of the Gulf Stream was called in question, just when meteorologists were disposed to assign to it effects of unusual importance. And in the course of the discussion whether there is in truth a Gulf Stream—or rather whether our shores are visited by a current which merits such a name—a variety of interesting facts were adduced, which were either before unknown or had attracted little attention. As at a recent meeting of the same society these doubts have been renewed, we propose to examine briefly, in the first place, a few of the considerations which have been urged against the existence of a current from the Gulf of Mexico to the neighborhood of our shores; and then, having rehabilitated the reputation of this celebrated ocean-river—as we believe we shall be able to do—we shall proceed to give a brief sketch of the processes by which the current-system of the North Atlantic is set and maintained in motion.

In reality the Gulf Stream is only a part of a

system of oceanic circulation; but in dealing with the arguments which have been urged against its very existence, we may confine our attention to the fact that, according to the views which had been accepted for more than a century, there is a stream of water which, running out of the Gulf Stream through the Narrows of Bemini, flows along the shores of the United States to Newfoundland, and thence right across the Atlantic to the shores of Great Britain. It is this last fact which is now called in question. The existence of a current as far as the neighborhood of Newfoundland is conceded, but the fact that the stream flows onward to our shores is denied.

The point on which the most stress is placed is the shallowness of the passage called the "Bemini Narrows," through which it is assumed that the whole of the Gulf current must pass. This passage has a width of about forty miles, and a depth of a little more than six hundred yards. The current which flows through it is perhaps little more than thirty miles in width, and a quarter of a mile in depth. It is asked with some appearance of reason, how this narrow current can be looked upon as the parent of that wide stream which is supposed to traverse the Atlantic with a mean width of some five or six hundred miles. Indeed, a much greater width has been assigned to it, though on mistaken grounds; for it has been remarked that since waifs and strays from the tropics are found upon the shores of Portugal, as well as upon those of Greenland, we must ascribe to the current a span equal to the enormous space separating these places. But the circumstance here dwelt upon can clearly be explained in another way. We know that of two pieces of wood thrown into the Thames at Richmond, one might be picked up at Putney, and the other at Gravesend. Yet we do not conclude that the width of the Thames is equal to the distance separating Putney from Gravesend. And doubtless the tropical waifs which have been picked up on the shores of Greenland and of Portugal have found their way thither by circuitous courses, and not by direct transmission along opposite edges of the great Gulf current.

But certainly the difficulty associated with the narrowness of the Bemini current is one deserving of careful attention. Are we free to identify a current six hundred miles in width with one which is but thirty miles wide, and not very deep? An increase of width certainly not less than thirtyfold would appear to correspond to a proportionate diminution of depth. And remembering that it is only near the middle of the Narrows that the Gulf Stream has a depth of four hundred yards, we could scarcely assign to the wide current in the mid-Atlantic a greater depth than ten or twelve yards. This depth seems altogether out of proportion to the enormous lateral extension of the current.

But besides that even this consideration would not suffice to disprove the existence of a current in the mid-Atlantic, an important circumstance remains to be mentioned. The current in the Narrows flows with great velocity—certainly not less than four or five miles an hour. As the current grows wider it flows more sedately; and opposite Cape Hatteras its velocity is already reduced to little more than three miles an hour. In the mid-Atlantic the current may be assumed to flow at a rate little exceeding a mile per hour, at the outside. Here, then, we have a circumstance which suffices to remove a large part of the difficulty arising from the narrowness of the Bemini current, and we can at once increase our estimate of the depth of the mid-Atlantic current fivefold.

But this is not all. It has long been understood that the current which passes out through the Narrows of Bemini corresponds to the portion of the great equatorial current which passes into the Gulf of Mexico between the West-Indian Islands. We cannot doubt that the barrier formed by those islands serves to divert a large portion of the equatorial current. The portion thus diverted finds its way, we may assume, along the outside of the West-Indian Archipelago, and thus joins the other portion—which has in the mean time made the circuit of the Gulf—as it issues from the Bemini Straits. All the maps in which the Atlantic currents are depicted present precisely such an outside current as we have here spoken of, and most of them assign to it a width exceeding that of the Bemini current. Indeed, were it not for the doubts which the recent discussions have thrown upon all the currents charted by seamen, we should have been content to point to this outside current as shown in the maps. As it is, we have thought it necessary to show that such a current must necessarily have an existence, since we cannot lose sight of the influence of the West-Indian Isles in partially damming up the passage along which the equatorial current would otherwise find its way into the Gulf of Mexico. Whatever portion of the great current is thus diverted must find a passage elsewhere, and no passage exists for it save along the outside of the West-Indian Isles.

The possibility that the wide current which has been assumed to traverse the mid-Atlantic may be associated with the waters which flow from the Gulf of Mexico, either through the Narrows or round the outside of the barrier formed by the West Indies, has thus been satisfactorily established. But we now have to consider difficulties which have been supposed to encounter our current on its passage from the Gulf to the mid-Atlantic.

Northward, along the shores of the United States, the current has been traced by the singular blueness of its waters until it has reached the neighborhood of

Newfoundland. Over a part of this course, indeed, the waters of the current are of indigo blue, and so clearly marked that their line of junction with the ordinary sea-water can be traced by the eye. "Often," says Captain Maury, "one half of a vessel may be perceived floating in Gulf-Stream water, while the other half is in common water of the sea—so sharp is the line, and such the want of affinity between the waters, and such, too, the reluctance, so to speak, on the part of those of the Gulf Stream to mingle with the littoral waters of the sea."

But it is now denied that there is any current beyond the neighborhood of Newfoundland — or that the warm temperature, which has characterized the waters of the current up to this point, can be detected farther out.

It is first noticed that, as the Gulf current must reach the neighborhood of Newfoundland with a northeasterly motion, and, if it ever reached the shores of the British Isles, would have to travel thither with an almost due easterly motion, there is a change of direction to be accounted for. This, however, is an old, and we had supposed exploded, fallacy. The course of the Gulf Stream from the Bemini Straits to the British Isles corresponds exactly with that which is due to the combined effects of the motion of the water and that of the earth upon its axis. Florida being much nearer than Ireland to the equator, has a much more rapid easterly motion. Therefore, as the current gets farther and farther north, the effect of the easterly motion thus imparted to it begins to show itself more and more, until the current is gradually changed from a northeasterly to an almost easterly stream. The process is the exact converse of that by which the air-currents from the north gradually change into the northwesterly trade-winds as they get farther south.

But it is further remarked that as the current passes out beyond the shelter of Newfoundland, it is impinged upon by those cold currents from the arctic seas which are known to be continually flowing out of Baffin's Bay and down the eastern shores of Greenland; and it is contended that these currents suffice, not merely to break up the Gulf current, but so to cool its waters that these could produce no effect upon the climate of Great Britain if they ever reached its neighborhood.

Here, again, we must remark that we are dealing with no new discovery. Captain Maury has already remarked upon this peculiarity. "At the very season of the year," he says, "when the Gulf Stream is rushing in greatest volume through the Straits of Florida, and hastening to the north with the greatest rapidity, there is a cold stream from Baffin's Bay, Labrador, and the coasts of the north, running south with equal velocity. . . . . One part of it underruns the Gulf

Stream, as is shown by the icebergs, which are carried in a direction tending across its course." There can be no doubt, in fact, that this last circumstance indicates the manner in which the main contest between the two currents is settled. A portion of the arctic current finds its way between the Gulf Stream and the continent of America; and this portion, though narrow, has a very remarkable effect in increasing the coldness of the American winters. But the main part, heavier, by reason of its coldness, than the surrounding water, sinks beneath the surface. And the well-known fact mentioned by Maury, that icebergs have been seen stemming the Gulf Stream, suffices to show how comparatively shallow that current is at this distance from its source, and thus aids to remove a difficulty which we have already had occasion to deal with.

Doubtless the cooling influence of the arctic currents is appreciable; but it would be a mistake to suppose that this influence can suffice to deprive the Gulf current of its distinctive warmth. If all the effect of the cold current were operative on the Gulf Stream alone, we might suppose that, despite the enormous quantity of comparatively warm water which is continually being carried northward, the current would be reduced to the temperature of the surrounding water. But this is not so. The arctic current not only cools the Gulf current, but the surrounding water also—possibly to a greater extent, for it is commonly supposed that a bed of ordinary sea-water separates the two main currents from each other. Thus the characteristic difference of temperature remains unaffected. But in reality we may assume that the cooling effect actually exercised by the arctic current upon the neighboring sea is altogether disproportionate to the immense amount of heat continually being carried northward by the Gulf Stream. It is astonishing how unreadily two sea-currents exchange their temperatures—to use a somewhat inexact mode of expression. The very fact that the littoral current of the United States is so cold—a fact thoroughly established—shows how little warmth this current has drawn from the neighboring seas. Another fact, mentioned by Captain Maury, bears in a very interesting manner upon this peculiarity. He says: "If any vessel will take up her position a little to the northward of Bermuda, and steering thence for the capes of Virginia, will try the water-thermometer all the way at short intervals, she will find its reading to be now higher, now lower; and the observer will discover that he has been crossing streak after streak of warm and cool water in regular alternations." Each portion maintains its own temperature even in the case of such warm streaks as these, all belonging to one current.

Similar considerations dispose of the arguments which have been founded on the temperature of the sea-bottom. It has been proved that the living creatures which people the lower depths of the sea exist under circumstances which evidence a perfect uniformity of temperature; and arguments on the subject of the Gulf Stream have been derived from the evidence of what is termed a minimum thermometer—that is, a thermometer which will indicate the lowest temperature it has been exposed to—let down into the depths of the sea. All such arguments, whether adduced against or in favor of the Gulf-Stream theory, must be held to be futile, since the thermometer in its descent may pass through several submarine currents of different temperature.

Lastly, an argument has been urged against the warming effects of the Gulf Stream upon our climate, which requires to be considered with some attention. It is urged that the warmth derived from so shallow a current as the Gulf Stream must be, by the time it has reached our shores, could not provide an amount of heat sufficient to affect our climate to any appreciable extent. The mere neighborhood of this water at a temperature slightly higher than that due to the latitude could not, it is urged, affect the temperature of the inland counties at all.

This argument is founded on a misapprehension of the beautiful arrangement by which Nature carries heat from one region to distribute it over another. Over the surface of the whole current the process of evaporation is going on at a greater rate than over the neighboring seas, because the waters of the current are warmer than those which surround them. The vapor thus rising above the Gulf Stream is presently wafted by the southwesterly winds to our shores and over our whole land. But as it thus reaches a region of comparative cold the vapor is condensed—that is, turned into fog, or mist, or cloud, according to circumstances. It is during this change that it gives out the heat it has brought with it from the Gulf Stream. For precisely as the evaporation of water is a process requiring heat, the change of vapor into water—whether in the form of fog, mist, cloud, or rain—is a process in which heat is given out. Thus it is that the southwesterly wind, the commonest wind we have, brings clouds and fogs and rain to us from the Gulf Stream, and with them brings the Gulf-Stream warmth.

Why the southwesterly winds should be so common, and how it is that over the Gulf Stream there is a sort of air-channel along which winds come to us as if by their natural pathway, we have not space here to inquire (see p. 183). The subject is full of interest, but it does not belong to the question we are considering.

It would seem that a mechanism involving the motion of such enormous masses of water as the current-system of the Atlantic should depend on the operation of very evident laws. Yet a variety of contradictory hypotheses have been put forward from time to time respecting this system of circulation, and even now the scientific world is divided between two opposing theories.

Of old the Mississippi River was supposed to be the parent of the Gulf Stream. It was noticed that the current flows at about the same rate as the Mississippi, and this fact was considered sufficient to support the strange theory that a river can give birth to an ocean-current.

It was easy, however, to overthrow this theory. Captain Livingston showed that the volume of water which is poured out of the Gulf of Mexico in the form of an ocean-stream is more than a thousand times greater than the volume poured into the Gulf by the Mississippi River.

Having overthrown this old theory of the Gulf Stream, Captain Livingston attempted to set up one which is equally unfounded. He ascribed the current to the sun's apparent yearly motion and the influence he exerts on the waters of the Atlantic. A sort of yearly tide is conceived, according to this theory, to be the true parent of the Gulf current. It need hardly be said, however, that a phenomenon which remains without change through the winter and summer seasons cannot possibly be referred to the operation of such a cause as a yearly tide.

It is to Dr. Franklin that we owe the first theory of the Gulf Stream which has met with general acceptance.

He held that the Gulf Stream is formed by the outflow of waters which have been forced into the Caribbean Sea by the trade-winds; so that the pressure of these winds on the Atlantic Ocean forms, according to Dr. Franklin, the true motive power of the Gulf-Stream machinery. According to Maury, this theory has "come to be the most generally-received opinion in the mind of seafaring people." It supplies a moving force of undoubted efficiency. We know that, as the trade-winds travel toward the equator, they lose their westerly motion. It is reasonable to suppose that this is caused by friction against the surface of the ocean, to which, therefore, a corresponding westerly motion must have been imparted.

There is a simplicity about Franklin's theory which commends it favorably to our consideration. But when we examine it somewhat more closely, several very decided flaws present themselves to our attention.

Consider, in the first place, the enormous mass of water moved by the supposed agency of the winds. Air has a weight—volume for volume—which is less than one eight-hundredth part of that of water. So that, to create a water-current, an air-current more than eight hundred times as large and of equal velocity must expend the whole of its motion. Now, the trade-winds are gentle winds, their velocity scarcely exceeding in general that of the more swiftly-moving portions of the Gulf Stream. But even assigning to them a velocity four times as great, we still want an air-current two hundred times as large as the water-current. And the former must give up the whole of its motion, which, in the case of so elastic a substance as air, would hardly happen, the upper air being unlikely to be much affected by the motion of the lower.

But this is far from being all. If the trade-winds blew throughout the year, we might be disposed to recognize their influence upon the Gulf Stream as a paramount, if not the sole one. But this is not the case. Captain Maury states that, "with the view of ascertaining the average number of days during the year that the northeast trade-winds of the Atlantic operate upon the currents between 25° north latitude and the equator, log-books containing no less than 380,284 observations on the force and direction of the wind in that ocean were examined. The data thus afforded were carefully compared and discussed. The results show that within these latitudes—and on the average—the wind from the northeast is in excess of the winds from the southwest only 111 days out of the 365. Now, can the northeast trades," he pertinently asks, "by blowing for less than one-third of the time, cause the Gulf Stream to run all the time, and without varying its velocity either to their force or to their prevalence?"

And besides this, we have to consider that no part of the Gulf Stream flows strictly before the trade-winds. Where the current flows most rapidly, namely, in the Narrows of Bemini, it sets against the wind, and for hundreds of miles after it enters the Atlantic, "it runs," says Maury, "right in the 'wind's eye.'" It must be remembered that a current of air directed with considerable force against the surface of still water has not the power of generating a current which can force its way far through the resisting fluid. If this were so, we might understand how the current, originating in sub-tropical regions, could force its way onward after the moving force had ceased to act upon it, and even carry the waters of the current right against the wind, after leaving the Gulf of Mexico. But experience is wholly opposed to this view. The most energetic currents are quickly dispersed when they reach a wide expanse of still water. For example, the Niagara below the falls is an immense and rapid river. Yet when it reaches Lake Ontario, "instead of preserving its character as a distinct and well-defined stream for several hundred miles, it spreads itself out, and its waters are immediately lost in those of the lake." Here, again, the question asked by Maury bears pertinently on the subject we are considering. "Why," he says, "should not the Gulf Stream do the same? It gradually enlarges itself, it is true; but, instead of mingling with the ocean by broad spreading, as the immense rivers descending into the northern lakes do, its waters, like a stream of oil in the ocean, preserve a distinctive character for more than three thousand miles."

The only other theory which has been considered in recent times to account satisfactorily for all the features of the Gulf-Stream mechanism was put forward, we believe, by Captain Maury. In this theory, the motive power of the whole system of oceanic circulation is held to be the action of the sun's heat upon the waters of the sea. We recognize two contrary effects as the immediate results of the sun's action. In the first place, by warming the equatorial waters, it tends to make them lighter; in the second place, by causing evaporation, it renders them salter, and so tends to make them heavier. We have to inquire which form of action is most effective. The inquiry would be somewhat difficult, if we had not the evidence of the sea itself to supply an answer. For it is an inquiry to which ordinary experimental processes would not be applicable. We must accept the fact that the heated water from the equatorial seas actually does float upon the cooler portions of the Atlantic, as evidence that the action of the sun results in making the water lighter.

Now, Maury says that the water thus lightened must flow over and form a surface-current toward the Poles; while the cold and heavy water from the polar seas, as soon as it reaches the temperate zone, must sink and form a submarine current. He recognizes in these facts the mainspring of the whole system of oceanic circulation. If a long trough be divided into two compartments, and we fill one with oil and the other with water, and then remove the dividing plate, we shall see the oil rushing over the water at one end of the trough, and the water rushing under the oil at the other. And if we further conceive that oil is continually being added at that end of the trough originally filled with oil, while water is continually added to the other, it is clear that the system of currents would continue in action: that is, there would be a continual flow of oil in one direction along the surface of the water, and of water in the contrary direction underneath the oil.

But Sir John Herschel maintains that no such effects as Maury describes could follow the action of the sun's heat upon the equatorial waters. He argues thus: Granting that these waters become lighter and expand in volume, yet they can only move upward, downward, or sideways. There can be nothing to cause either of the first two forms of motion; and as for motion sideways, it can only result from the gradual slope caused by the bulging of the equatorial waters. He proceeds to show that this slope is so slight that we cannot look upon it as competent to form any sensible current from the equatorial toward the polar seas. And even if it could, he says, the water thus flowing off would have an eastward instead of a westward motion, precisely as the counter-trade-winds, blowing from equatorial to polar regions, have an eastward motion.

It is singular how completely the supporter of each rival view has succeeded in overthrowing the arguments of his opponent. Certainly Maury has shown with complete success that the inconstant trade-winds cannot account for the constant Gulf current, which does not even flow before them, but, in places, exactly against their force. And the reasoning of Sir John Herschel seems equally cogent, for certainly the flow of water from equatorial toward polar regions ought from the first to have an eastward, instead of a westward motion; whereas the equatorial current, of which the Gulf Stream is but the continuation, flows from east to west, right across the Atlantic.

Equally strange is it to find that each of these eminent men, having read the arguments of the other, reasserts, but does not effectually defend, his own theory, and repeats with even more damaging effect his arguments against the rival view.

Yet one or other theory must at least point to the true view, for the Atlantic is subject to no other agencies which can for a moment be held to account for a phenomenon of such magnificence as the Gulf Stream.

It appears to us that, on a close examination of the Gulf-Stream mechanism, the true mainspring of its motion can be recognized. Compelled to reject the theory that the trade-winds generate the equatorial current westward, let us consider whether Herschel's arguments against the "heat-theory" may not suggest a hint for our guidance. He points out that an overflow from the equator poleward would result in an eastward, and not in a westward current. This is true. It is equally true that a flow of water toward the equator would result in a westward current. But no such flow is observed. Is it possible that there may be such a flow, but that it takes place in a hidden manner? Clearly there may be. Sub-marine currents toward the equator would have precisely the kind of motion we require, and, if any cause drew them to the surface near the equator, they would account in full for the great equatorial westward current.

At this point we begin to see that an important circumstance has been lost sight of in dealing with the heat-theory. The action of the sun on the surface-water of the equatorial Atlantic has only been considered with reference to its warming effects. But we must not forget that this action has drying effects also. It evaporates enormous quantities of water, and we have to inquire whence the water comes by which the sea-level is maintained. A surface-flow from the subtropical seas would suffice for this purpose, but no such flow is observed. Whence, then, can the water come but from below? Thus we recognize the fact that a process resembling suction is continually taking place over the whole area of the equatorial Atlantic, the agent being the intense heat of the tropical sun. No one can doubt that this agent is one of adequate power. Indeed, the winds, conceived by Franklin to be the primary cause of the Atlantic currents, are in reality due to the merest fraction of the energy inherent in the sun's heat.

We have other evidence that the indraught is from below, in the comparative coldness of the equatorial current. The Gulf Stream is warm by comparison with the surrounding waters, but the equatorial current is cooler than the tropical seas. According to Professor Ansted, the southern portion of the equatorial current, as it flows past Brazil, "is everywhere a cold current, generally from four to six degrees below the adjacent ocean."

Having once detected the mainspring of the Gulf-Stream mechanism, or rather of the whole system of oceanic circulation—for the movements observed in the Atlantic have their exact counterpart in the Pacific—we have no difficulty in accounting for all the motions which that mechanism exhibits. We need no longer look upon the Gulf Stream as the rebound of the equatorial current from the shores of North America. Knowing that there is an underflow toward the equator, we see that there must be a surface-flow toward the Poles. And this flow must as inevitably result in an easterly motion, as the underflow toward the equator results in a westerly motion. We have, indeed, the phenomena of the trades and counter-trades exhibited in water-currents instead of air-currents.

(From *St. Paul's*, September, 1869.)*

## *FLOODS IN SWITZERLAND.*

DURING the past few weeks we have witnessed a succession of remarkable evidences of Nature's destructive powers. The fires of Vesuvius, the earth-throes of the sub-equatorial Andes, and the submarine disturbance which has shaken Hawaii, have presented to us the various forms of destructive action which the earth's subterranean forces can assume. In the disastrous floods which have recently visited the Alpine cantons of Switzerland, we have evidence of the fact that natural forces which we are in the habit of regarding as beneficent and restorative may exhibit themselves as agents of the most wide-spread destruction. We have pointed out elsewhere (see p. 249) how enormous is the amount of power of which the rain-cloud is the representative; and in doing so we have endeavored to exhibit the contrast between the steady action of the falling shower and the energy of the processes of which rain is in reality the equivalent. But in the floods which have lately ravaged Switzerland we see the same facts illustrated, not by numerical calculations or by the results of philosophical experiments,

* See also *The Student* for July, 1868.

but in action, and that action taking place on the most widely-extended scale. The whole of the southeastern, or, as it may be termed, the Alpine half of Switzerland, has suffered from these floods. If a line be drawn from the Lake of Constance, in the northeast of Switzerland, to the Col de Balme, in the southwest, it will divide Switzerland into two nearly equal portions, and scarcely a canton within the eastern of these divisions has escaped without great damage.

The cantons which have suffered most terribly are those of Tessin, Grisons, and St. Gall. The St. Gothard, Splugen, and St. Bernhardin routes, have been rendered impassable. Twenty-seven lives were lost in the St. Gothard Pass, besides horses and wagons full of merchandise. It is stated that on the three routes upward of eighty persons perished. In the village of Loderio alone, no less than fifty deaths occurred. So terrible a flood has not taken place since the year 1834. Nor have the cantons of Uri and Valais escaped. From Unterwalden we hear that the heavy rains which took place a fortnight ago have carried away several large bridges, and many of the rivers continue still very swollen. We have already described how enormous the material losses are which have been caused by these floods. Many places are under water; others in ruins or absolutely destroyed. In Tessin alone the damage is estimated at forty thousand pounds sterling.

A country like Switzerland must always be liable to the occurrence, from time to time, of catastrophes of this sort. Or rather, perhaps, we should draw a distinction between the two divisions of Switzerland referred to above. Of these the one may be termed the mountain half, and the other the lake half of the country. It is the former portion of the country which is principally subject to the dynamical action of water. A long-continued and heavy rainfall over the higher lands cannot fail to produce a variety of remarkable effects, where the arrangement of mountains and passes, hills, valleys, and ravines, is so complicated. There are places where a large volume of water can accumulate until the barriers which have opposed its passage to the plains burst under its increasing weight; and then follow those destructive rushes of water which sweep away whole villages at once. It is, in fact, the capacity of the Swiss mountain-region for damming up water, far more than any other circumstance, which renders the Swiss floods so destructive.

And then it must be remembered that there are at all times suspended over the plains and valleys which lie beneath the Alpine ranges enormous masses of water in the form of snow and ice. Although in general these suffer no changes but those due to the partial melting which takes place in summer, and the renewed accumulation which takes place in winter, yet when heavy rains fall upon the less elevated portions of the Alpine snow, they not only melt that snow much more rapidly than the summer sun would do, but they wash down large masses, which add largely to the destructive power of the descending waters.

The most destructive floods which have occurred in Switzerland have usually been those which take place in early summer. The floods which inundated the plains of Martigny in 1818 were a remarkable instance of the effects which result from the natural damming up of large volumes of water in the upper parts of the Alpine hill-country. The whole of the valley of Bagnes, one of the largest of the lateral branches of the main valley of the Rhone above Geneva, was converted into a lake, in the spring of 1818, by the damming up of a narrow pass into which avalanches of snow and ice had been precipitated from a lofty glacier overhanging the bed of the river Dranse. The icy barrier enclosed a lake no less than half a league in length and an eighth of a mile wide, and in places two hundred feet deep. The inhabitants of the neighboring villages were terrified by the danger which was to be apprehended from the bursting of the barrier. They cut a gallery seven hundred feet long through the ice, while the waters had as yet risen to but a moderate height; and when the waters began to flow through this channel, its course was deepened by the melting of the ice, and at length nearly half the contents of the lake were safely carried off. It was hoped that the process would continue, and the country be saved from the danger which had been so long impending over it. But as the heat of the weather increased, the central part of the barrier slowly melted away, until it became too weak to bear the enormous weight of water which was pressing against it. At length it gave way, so suddenly and completely that all the water which remained in the lake rushed out in half an hour. The downward passage of the water illustrated, in a very remarkable way, the fact that the chief mischief of floods is occasioned where water is checked in its outflow. For it is related that, "in the course of their descent the waters encountered several narrow gorges, and at each of these they rose to a great height, and then burst with new violence into the next basin, sweeping along forests, houses, bridges, and cultivated land." Along the greater part of its course the flood resembled rather a moving mass of rock and mud than a stream of water. Enormous masses of granite were torn out of the sides of the valleys and whirled for hundreds of yards along the course of the flood. M. Escher relates that one of the fragments thus swept along was no less than sixty yards in circumference. At first the water rushed onward at a rate of more than a mile in three minutes, and the whole distance (forty-five miles) which separates the valley of Bagnes from the Lake of Geneva was traversed in little more than six hours. The bodies of persons who had been drowned in Martigny were found floating on the farther side of the Lake of Geneva, near Vevey. Thousands of trees were torn up by the roots, and the ruins of buildings which had been overthrown by the flood were carried down beyond Martigny. In fact, the flood at this point was so high that some of the houses in Martigny "were filled with mud up to the second story." Beyond Martigny the flood did but little damage, as it here expanded over the plain, and was at once reduced in depth and velocity.

(From the *Daily News* for October 20, 1868.)

---

## *A GREAT TIDAL WAVE.*

DURING the last few days anxious questionings have been heard respecting the next spring tides. A certain naval officer, who conceives that he can trace in the relative positions of the sun and moon the secret of every important change of weather, has described in the columns of a contemporary the threatening significance of the approaching conjunction of the sun and moon. He predicts violent atmospheric disturbances; though in another place he tells us merely that the conjunction is to cause "unsettled weather," a state of matters to which we in England have become tolerably well accustomed.

But people are asking what is the actual relation which is to bring about such terrible events. The matter is very simple. On October 5th, the moon will be new—in other words, if it were not for the brightness of the sun, we should see the moon close by that luminary on the heavens. Thus the sun and moon will pull with combined effect upon the waters of the earth, and so cause what are called spring tides. This, of course, happens at the time of every new moon. But sometimes the moon exerts a more effective pull than at other times; and the same happens also in the case of the sun; and on October 5th, it happens that both the sun and the moon will give a particularly vigorous haul upon the earth's waters. As regards the sun, there is nothing unusual. Every October his pull on the ocean is much the same as in preceding Octobers. But October is a month of high solar tides—and for these reasons: In September, as every one knows, the sun crosses the equinoctial; and other things being equal it would be when on the equinoctial that his power to raise a tidal wave would be greatest. But other things are not equal; for the sun is not always at a fixed distance from the earth. He is nearest in January; so that he would exert more power in that month than in any other if his force depended solely on distance. As matters actually stand, it will be obvious that at some time between September and January the sun's tidal power would have a maximum value. Thus it is that October is a month of high solar tidal waves.

But is it the lunar wave which will be most effectively strengthened at the next spring tide. If we could watch the lunar tidal wave alone (instead of always finding it combined with the solar wave) we should find it gradually increasing, and then gradually diminishing, in a period of about a lunar month. And we should find that it was always largest when the moon looked largest, and *vice versa*. In other words, when the moon is in perigee the lunar wave is largest. But then there is another consideration. The lunar wave would vary according to the moon's proximity to the equinoctial; and (other things being equal) would be largest when the moon is exactly opposite the earth's equator. If the two effects are combined, that is, if the moon happens to be in perigee and on the equinoctial at the same time, then of course we get the largest lunar tidal wave we can possibly have.

Now, this "largest lunar wave" occurs at somewhat long intervals, because the relation on which it depends is one which is, so to speak, exceptional. Still the relation does recur, and with a certain degree of regularity. When it happens, however, it by no means follows that we have a very high tide; because it may occur when the tides are near "neap;" in other words, when the sun and moon exert opposing effects. The largest lunar wave cannot stand the drain which the solar wave exerts upon it at the time of neap tides. Nor would the large lunar tidal wave produce an exceptionally high tide, even though it were not the time of "neap," or were tolerably near the time of "spring" tides. Only when it happens that a large lunar wave combines fully with the solar wave do we get very high tides. And when, in addition to this relation, we have the solar wave nearly at a maximum, we get the highest of all possible tides. This is what will happen, or all but happen, on October 5th next. The combination of circumstances is almost the most effective that can possibly exist.

But, after all, high tides depend very importantly on other considerations than astronomical ones. Most of us remember how a predicted high tide some two years ago turned out to be very moderate, or, if we may use the expression, a very "one-horse" affair indeed, because the winds had not been consulted, and exerted their influence against the astronomers. A long succession of winds blowing off-shore would reduce a spring tide to a height scarcely exceeding the ordinary neap. On the other hand, if we should have a long succession of westerly winds from the Atlantic before the approaching high tide, it is certain that a large amount of mischief may be done in some of our river-side regions.*

As for the predicted weather changes, they may be

* The wave did little mischief.

regarded as mere moonshine. A number of predictions, founded on the motions of the sun and moon, have found a place during many months past in the columns of a contemporary; but there has been no greater agreement between these predictions and the weather actually experienced than any one could trace between Old More's weather prophecies and recorded weather changes. In other words, there have been certain accordances which would be very remarkable indeed if they did not happen to be associated with as many equally remarkable discordances. Random predictions would be quite as satisfactory.

A very amusing misprint has found its way into many newspapers in connection with the coming tide. It is interesting as serving to show how little is really known by the general public about some of the simplest scientific matters. The original statement announced that the sun would not be in perihelion by so many seconds of semi-diameter, in itself a very incorrect mode of expression. Still it was clear that what was meant was, that the earth would be so far from the place of nearest approach to the sun that the latter would not look as large as it possibly can by so many seconds of semi-diameter. In many papers, however, we read that the "sun will not be in perihelion by so many seconds of mean chronometer!" Who first devised this marvellous reading is unknown—he should have a statue.

(From the *Daily News* for September 27, 1869.)

## *DEEP-SEA DREDGINGS.*

Men have ever been strangely charmed by the unknown and the seemingly inaccessible. The astronomer exhibits the influence of this charm as he constructs larger and larger telescopes, that he may penetrate more and more deeply beyond the veil which conceals the greater part of the universe from the unaided eye. The geologist seeking to piece together the fragmentary records of the past which the earth's surface presents to him, is equally influenced by the charm of mystery and difficulty. And the microscopist who tries to force from Nature the secret of the infinitely little, is led on by the same strange desire to discover just those matters which Nature has been most careful to conceal from us.

The energy with which in recent times men have sought to master the problem of deep-sea sounding and deep-sea dredging is, perhaps, one of the most striking instances ever afforded of the charm which the unknown possesses for mankind. Not long ago, one of the most eminent geographers of the sea spoke regretfully about the small knowledge men have obtained of the depths of ocean. "Greater difficulties," he remarked, "than any presented by the problem of deep-sea research have been overcome in other branches of physical inquiry. Astronomers have measured the volumes and weighed the masses of the most distant planets, and increased thereby the stock of human knowledge. Is it creditable to the age that the depths of the sea should remain in the category of unsolved problems? that its 'ooze and bottom' should be a sealed volume, rich with ancient and eloquent legends, and suggestive of many an instructive lesson that might be useful and profitable to man?"

Since that time, however, deep-sea dredging has gradually become more and more thoroughly understood and mastered. Recently, when the telegraphic cable which had lain so many months at the bottom of the Atlantic was hauled on board the "Great Eastern" from enormous depths, men were surprised and almost startled by the narrative. The appearance of the ooze-covered cable as it was slowly raised toward the surface, and the strange thrill which ran through those who saw it and remembered through what mysterious depths it had twice passed; its breaking away almost from the very hands of those who sought to draw it on board; and the successful renewal of the attempt to recover the cable—all these things were heard of as one listens to a half-incredible tale. Yet when that work was accomplished deep-sea dredging had already been some time a science, and many things had been achieved by its professors which presented, in reality, greater practical difficulties than the recovery of the Atlantic Cable.

Recently, however, deep-sea researches have been carried on with results which are even more sensational, so to speak, than the grappling feat which so surprised us. Seas so deep that many of the loftiest summits of the Alps might be completely buried beneath them have been explored. Dredges weighing with their load of mud nearly half a ton have been hauled up without a hitch from depths of some 14,000 feet. But not merely has comparatively rough work of this sort been achieved, but by a variety of ingenious contrivances men of science have been able to measure the temperature of the sea at depths where the pressure is so enormous as to be equivalent to a weight of more than 430 tons on every square foot ot surface.

The results of these researches are even more remarkable and surprising, however, than the means by which they have been obtained. Sir Charles Lyell has fairly spoken of them as so astonishing "that they have to the geologist almost a revolutionary character." Let us consider a few of them.

No light can be supposed to penetrate to the enormous depths just spoken of. Therefore, how certainly we might conclude that there can be no life there! If, instead of dealing with the habitability of planets, Whewell, in his "Plurality of Worlds," had been considering the question whether at depths of two or three miles living creatures could subsist, how convincingly would he have proved the absurdity of such a supposition! Intense cold, perfect darkness, and a persistent pressure of two or three tons to the square inch—such, he might have argued, are the conditions under which life exists, if at all, in those dismal depths. And even if he had been disposed to concede the bare possibility that life of some sort may be found there, then certainly he would have urged, some new sense must replace sight—the creatures in these depths can assuredly have no eyes, or only rudimentary ones.

But the recent deep sea-dredgings have proved that not only does life exist in the very deepest parts of the Atlantic, but that the beings which live and move and have their being beneath the three-mile mountain of water have eyes which the ablest naturalists pronounce to be perfectly developed. Light, then, of some sort must exist in those abysms, though whether the home of the deep-sea animals be phosphorescent, as Sir Charles Lyell suggests, or how light may reach these creatures, we have no present means of determining.

If there is one theory which geologists have thought more justly founded than all others, it is the view that the various strata of the earth were formed at different times. A chalk district, for example, lying side by side with a sandstone district, has been referred to a totally different era. Whether the chalk was formed first, or whether the sandstone existed before the minute races came into being which formed the cretaceous stratum, might be a question. But no doubt existed in the minds of geologists that each formation belonged to a distinct period. Now, however, Dr. Carpenter and Professor Thomson may fairly say, "We have changed all this." It has been found that at points of the sea-bottom only eight or ten miles apart, there may be in progress the formation of a cretaceous deposit and of a sandstone region, each with its own proper fauna. "Wherever similar conditions are found upon the dry land of the present day," remarks Dr. Carpenter, "it has been supposed that the formation of chalk and the formation of sandstone must have been separated from each other by long periods, and the discovery that they may actually coexist upon adjacent surfaces has done no less than strike at the very root of the customary assumptions with regard to geological time." *

Even more interesting, perhaps, to many, are the results which have been obtained respecting the varying temperatures of deep-sea regions. The peculiarity just considered is, indeed, a consequence of such variations; but the fact itself is at least as interesting as the consequences which flow from it. It throws light

* This opinion Dr. Carpenter has since somewhat modified. It will be remembered, of course, that the evidence derived from the nature of superposed strata is in no way affected by what is shown above to hold as respects adjacent deposits.

on the long-standing controversy respecting the oceanic circulation. It has been found that the depths of the equatorial and tropical seas are colder than those of the North Atlantic. In the tropics the deep-sea temperature is considerably below the freezing-point of fresh water; in the deepest part of the Bay of Biscay the temperature is several degrees above the freezing-point. Thus one learns that the greater part of the water which lies deep below the surface of the equatorial and tropical seas comes from the Antarctic regions, though undoubtedly there are certain relatively narrow currents which carry the waters of the Arctic seas to the tropics. The great point to notice is that the water under the equatorial seas must really have travelled from polar regions. A cold of 30° can be explained in no other way. We see at once, therefore, the explanation of those westerly equatorial currents which have been so long a subject of contest. Sir John Herschel failed to prove that they are due to the trade-winds, but Maury failed equally to prove that they are due to the great warmth and consequent buoyancy of the equatorial waters. In fact, while Maury showed very convincingly that the great system of oceanic circulation is carried on *despite* the winds, Herschel proved in an equally convincing manner that the overflow conceived by Maury should result in an easterly instead of a westerly current. Recently the theory was put forward that the continual process of evaporation going on in the equatorial regions leads to an indraught of cold water in bottom-currents from the polar seas. Such currents coming *toward* the equator, that is, travelling from latitudes where the earth's eastwardly motion is less to latitudes in which that motion is greater, would lag behind, that is, would have a westwardly motion. It seems now placed beyond a doubt that this is the true explanation of the equatorial ocean-currents.

Such are a few, and but a few, among the many interesting results which have followed from the recent researches of Dr. Carpenter and Professor Thomson into the hitherto little known depths of the great sea.

(From the *Spectator*, December 4, 1869.)

---

## *THE TUNNEL THROUGH MONT CENIS.*

Men flash their messages across mighty continents and beneath the bosom of the wide Atlantic; they weigh the distant planets, and analyze the sun and the stars; they span Niagara with a railway bridge, and pierce the Alps with a railway tunnel: yet the poet of the age in which all these things are done or doing sings, "We men are a puny race." And certainly, the great works which belong to man as a race can no more be held to evidence the importance of the individual man than the vast coral reefs and atolls of the

Pacific can be held to evidence the working-power of the individual coral polype. But if man, standing alone, is weak, man working according to the law assigned to his race from the beginning—that is, in fellowship with his kind—is verily a being of power.

Perhaps no work ever undertaken by men strikes one as more daring than the attempt to pierce the Alps with a tunnel. Nature seems to have upreared these mighty barriers as if with the design of showing man how weak he is in her presence. Even the armies of Hannibal and Napoleon seemed all but powerless in the face of these vast natural fastnesses. Compelled to creep slowly and cautiously along the difficult and narrow ways which alone were open to them, decimated by the chilling blasts which swept the face of the rugged mountain-range, and dreading at every moment the pitiless swoop of the avalanche, the French and Carthaginian troops exhibited little of the pomp and dignity which we are apt to associate with the operations of warlike armies. Had the denizen of some other planet been able to watch their progress, he might indeed have said, " These men are a puny race." In this only, that they succeeded, did the troops of Hannibal and Napoleon assert the dignity of the human race. Grand as was the aspect of Nature, and mean as was that of man during the progress of the contest, it was Nature that was conquered—man that overcame. And now man has entered on a new conflict with Nature in the gloomy fastnesses of the Alps. The barrier which he had scaled of old he has now undertaken to pierce. And the work—bold and daring as it seems—is three parts finished.

The Mont Cenis tunnel was sanctioned by the Sardinian Government in 1857, and arrangements were made for fixing the perforating machinery in the years 1858 and 1859. But the work was not actually commenced until November, 1860. The tunnel, which will be fully seven and a half miles in length, was to be completed in twenty-five years. The entrance to the tunnel on the side of France is near the little village of Fourneau, and lies 3,946 feet above the level of the sea. The entrance on the side of Italy is in a deep valley at Bardonèche, and lies 4,380 feet above the sea-level. Thus there is a difference of level of 434 feet. But the tunnel will actually rise 445 feet above the level of the French end, attaining this height at a distance of about four miles from that extremity; in the remaining three and three-quarter miles there will be a fall of only ten feet, so that this part of the line will be practically level.

The rocks through which the excavations have been made have been for the most part very difficult to work. Those who imagine that the great mass of our mountain-ranges consists of such granite as is made use of in our buildings, and is uniform in texture and hardness, greatly underrate the difficulties with which the engineers of this gigantic work have had to contend. A large part of the rock consists of a crystallized calcareous schist, much broken and contorted; and through this rock run in every direction large masses of pure quartz. It will be conceived how difficult the work has been of piercing through so diversified a substance as this. The perforating machines are calculated to work best when the resistance is uniform; and it has often happened that the unequal resistance offered to the perforators has resulted in injury to the chisels. But before the work of perforating began, enormous difficulties had to be contended with. It will be understood that, in a tunnel of such vast length, it was absolutely necessary that the perforating processes carried on from the two ends should be directed with the most perfect accuracy. It has often happened in short tunnels that a want of perfect coincidence has existed between the two halves of the work, and the tunnellers from one end have sometimes altogether failed to meet those from the other. But in a short tunnel this want of coincidence is not very important, because the two interior ends of the tunnellings cannot in any case be far removed from each other. But in the case of the Mont Cenis tunnel any inaccuracy in the direction of the two tunnellings would have been fatal to the success of the work, since when the two ought to meet it might be found that they were laterally separated by two or three hundred yards. Hence it was necessary, before the work began, to survey the intermediate country, so as to ascertain with the most perfect accuracy the bearings of one end of the tunnel from the other. "It was necessary," says the narrative of these initial labors, "to prepare accurate plans and sections for the determination of the levels, to fix the axis of the tunnel, and to 'set it out' on the mountain-top; to erect observatories and guiding-signals, solid, substantial, and true." When we remember the nature of the passes over the Cenis, we can conceive the difficulty of setting out a line of this sort over the Alpine range. The necessity of continually climbing over rocks, ravines, and precipices in passing from station to station involved difficulties which, great as they were, were as nothing when compared with the difficulties resulting from the bitter weather experienced on those rugged mountain-heights. The tempests which sweep the Alpine passes—the ever-recurring storms of rain, sleet, and driving snow, are trying to the ordinary traveller. It will be understood, therefore, how terribly they must have interfered with the delicate processes involved in surveying. It often happened that for days together no work of any sort could be done owing to the impossibility of using levels and theodolites when exposed to the stormy weather and bitter cold of these lofty passes. At length, however, the work was completed, and that with such success that the greatest deviation from exactitude was less than a single foot for the whole length of the seven and a half miles.

Equally remarkable and extensive were the labors connected with the preparatory works. New and solid roads, bridges, canals, magazines, workshops, forges, furnaces, and machinery, had to be constructed; residences had to be built for the men, and offices for the engineers; in fact, at each extremity of the tunnel a complete establishment had to be formed. Those who have traversed Mont Cenis since the works began have been perplexed by the strange appearance and character of the machinery and establishments to be seen at Modane and Fourneau. The mass of pipes and tubes, tanks, reservoirs, and machinery, which would be marvellous anywhere, has a still stranger look in a wild and rugged Alpine pass.

(From the *Daily News.*)

---

## *TORNADOES.*

The inhabitants of the earth are subjected to agencies which—beneficial doubtless in the long-run, perhaps necessary to the very existence of terrestrial races—appear, at first sight, energetically destructive. Such are—in order of destructiveness—the hurricane, the earthquake, the volcano, and the thunder-storm. When we read of earthquakes such as those which overthrew Lisbon, Callao, and Riobamba, and learn that one hundred thousand persons fell victims in the great Sicilian earthquake in 1693, and probably three hundred thousand in the two earthquakes which assailed Antioch in the years 526 and 612, we are disposed to assign at once to this devasting phenomenon the foremost place among the agents of destruction. But this judgment must be reversed when we consider that earthquakes—though so fearfully and suddenly destructive both to life and property—yet occur but seldom compared with wind-storms, while the effects of a real hurricane are scarcely less destructive than those of the sharpest shocks of earthquake. After ordinary storms long miles of the sea-coast are strewn with the wrecks of many once gallant ships, and with the bodies of their hapless crews. In the spring of 1866 there might be seen at a single view from the heights near Plymouth twenty-two shipwrecked vessels, and this after a storm, which, though severe, was but trifling compared with the hurricanes which sweep over the torrid zones, and thence—scarcely diminished in force—as far north sometimes as our own latitudes. It was in such a hurricane that the "Royal Charter" was wrecked, and hundreds of stout ships with her. In the great hurricane of 1780, which commenced at Barbadoes and swept across the whole breadth of the North Atlantic, fifty sail were driven ashore at the Bermudas, two line-of-battle ships went down at sea, and upward of twenty thousand persons lost their lives on the land. So tremendous was the force of this hurricane (Captain Maury tells us) that "the bark was blown from the trees, and the fruits of the earth destroyed; the very bottom and depths of the sea were uprooted —forts and castles were washed away, and their great guns carried in the air like chaff; houses were razed; ships wrecked; and the bodies of men and beasts lifted up in the air and dashed to pieces in the storm" —an account, however, which (though doubtless faithfully rendered by Maury from the authorities he consulted) must perhaps be accepted *cum grano*, and especially with reference to the great guns carried in the air "like chaff." *

In the gale of August, 1782, all the trophies of Lord Rodney's victory, except the "Ardent," were destroyed, two British ships-of-the-line foundered at sea, numbers of merchantmen under Admiral Graves's convoy were wrecked, and at sea alone three thousand lives were lost.

But quite recently a storm far more destructive than these swept over the Bay of Bengal. Most of our readers doubtless remember the great gale of October, 1864, in which all the ships in harbor at Calcutta were swept from their anchorage, and driven one upon another in inextricable confusion. Fearful as was the

* We remember to have read that in this hurricane guns which had long lain under water were washed up like mere drift upon the beach. Perhaps this circumstance grew gradually into the incredible story above recorded.

loss of life and property in Calcutta harbor, the destruction on land was greater. A vast wave swept for miles over the surrounding country, embankments were destroyed, and whole villages, with their inhabitants, were swept away. Fifty thousand souls, it is believed, perished in this fearful hurricane.

The gale which has just ravaged the Gulf of Mexico adds another to the long list of disastrous hurricanes. As we write, the effects produced by this tornado are beginning to be made known. Already its destructiveness has become but too certainly evidenced.

The laws which appear to regulate the generation and the progress of cyclonic storms are well worthy of careful study.

The regions chiefly infested by hurricanes are the West Indies, the southern parts of the Indian Ocean, the Bay of Bengal, and the China Seas. Each region has its special hurricane season.

In the West Indies, cyclones occur principally in August and September, when the southeast monsoons are at their height. At the same season the African southwesterly monsoons are blowing. Accordingly, there are two sets of winds, both blowing heavily and steadily from the Atlantic, disturbing the atmospheric equilibrium, and thus in all probability generating the great West-Indian hurricanes. The storms thus arising show their force first at a distance of about six or seven hundred miles from the equator, and far to the east of the region in which they attain their greatest fury. They sweep with a northwesterly course to the Gulf of Mexico, pass thence northward, and so to the northeast, sweeping in a wide curve (resembling the letter U placed thus ⊂) around the West-Indian seas, and thence travelling across the Atlantic, generally expending their fury before they reach the shores of Western Europe. This course is the storm-track (or storm-⊂ as we shall call it). Of the behavior of the winds as they traverse this track, we shall have to speak when we come to consider the peculiarity from which these storms derive their names of "cyclones" and "tornadoes."

The hurricanes of the Indian Ocean occur at the "changing of the monsoons." "During the interregnum," writes Maury, "the fiends of the storm hold their terrific sway." Becalmed often for a day or two, seamen hear moaning sounds in the air, forewarning them of the coming storm. Then, suddenly, the winds break loose from the forces which have for a while controlled them, and "seem to rage with a fury that would break up the fountains of the deep."

In the North Indian seas hurricanes rage at the same season as in the West Indies.

In the China seas occur those fearful gales known among sailors as "typhoons," or "white squalls." These take place at the changing of the monsoons.

Generated, like the West-Indian hurricanes, at a distance of some ten or twelve degrees from the equator, typhoons sweep—in a curve similar to that followed by the Atlantic storms—around the East-Indian Archipelago, and the shores of China to the Japanese Islands.

There occur land-storms, also, of a cyclonic character in the valley of the Mississippi. "I have often observed the paths of such storms," says Maury, "through the forests of the Mississippi. There the track of these tornadoes is called a 'wind-road,' because they make an avenue through the wood straight along, and as clear of trees as if the old denizens of the forest had been cleared with an axe. I have seen trees three or four feet in diameter torn up by the roots, and the top, with its limbs, lying next the hole whence the root came." Another writer, who was an eye-witness to the progress of one of these American land-storms, thus speaks of its destructive effects: "I saw, to my great astonishment, that the noblest trees of the forest were falling into pieces. A mass of branches, twigs, foliage, and dust, moved through the air, whirled onward like a cloud of feathers, and passing, disclosed a wide space filled with broken trees, naked stumps, and heaps of shapeless ruins, which marked the path of the tempest."

If it appeared, on a careful comparison of observations made in different places, that these winds swept directly along those tracks which they appear to follow, a comparatively simple problem would be presented to the meteorologist. But this is not found to be the case. At one part of a hurricane's course the storm appears to be travelling with fearful fury along the true storm-⊏; at another less furiously directly across the storm-track; at another, but with yet diminished force, though still fiercely, in a direction exactly opposite to that of the storm-track.

All these motions appear to be fairly accounted for by the theory that the true path of the storm is a spiral—or rather, that while the centre of disturbance continually travels onward in a widely-extended curve, the storm-wind sweeps continually around the centre of disturbance, as a whirlpool around its vortex.

And here a remarkable circumstance attracts our notice, the consideration of which points to the mode in which cyclones may be conceived to be generated. It is found, by a careful study of different observations made upon the same storm, that cyclones in the northern hemisphere *invariably* sweep round the onward travelling vortex of disturbance in *one* direction, and southern cyclones in the contrary direction. If we place a watch, face upward, upon one of the northern cyclone regions in a Mercator's chart, then the motion of the hands is *contrary* to the direction in which the cyclone whirls; when the watch is shifted to a southern cyclone region, the motion of the hands takes place in the same direction as the cyclone motion. This peculiarity is converted into the following rule-of-thumb for sailors who encounter a cyclone, and seek to escape from the region of fiercest storm: *Facing the wind, the centre or vortex of the storm lies to the right in the northern, to the left in the southern, hemisphere.* Safety lies in flying from the centre in every case save one—that is, when the sailor lies in the direct track of the advancing vortex. In this case, to fly from the centre would be to keep in the storm-track; the proper course for the sailor when thus situated is to steer for the calmer side of the storm-track. This is always the outside of the ⊏, as will appear from a moment's consideration of the spiral curve traced out by a cyclone. Thus, if the seaman *scud before the wind*—in all other cases a dangerous expedient in a cyclone*—he will probably escape unscathed. There is, however, this danger, that the storm-track may extend to or even slightly overlap the land, in which case scudding before the gale would bring the ship upon a lee-shore. And in this way many gallant ships have, doubtless, suffered wreck.

The danger of the sailor is obviously greater, however, when he is overtaken by the storm on the inner side of the storm-⊏. Here he has to encounter the

* A ship by scudding before the gale may—if the captain is not familiar with the laws of cyclones—go *round and round* without escaping. The ship "Charles Heddle" did this in the East Indies, going round no less than *five times.*

double force of the cyclonic whirl and of the advancing storm-system, instead of the difference of the two motions, as on the outer side of the storm-track. His chance of escape will depend on his distance from the central path of the cyclone. If near to this, it is equally dangerous for him to attempt to scud to the safer side of the track, or to beat against the wind by the shorter course which would lead him out of the storm-[illegible] on its inner side. It has been shown by Colonel Sir W. Reid that this is the quarter in which vessels have been most frequently lost.

But even the danger of this most dangerous quarter admits of degrees. It is greatest where the storm is sweeping round the most curved part of its track, which happens in about latitude twenty-five or thirty degrees. In this case, a ship may pass twice through the vortex of the storm. Here hurricanes have worked their most destructive effects. And thus it happens that sailors dread, most of all, the part of the Atlantic near Florida and the Bahamas, and the region of the Indian Ocean which lies south of Bourbon and Mauritius.

To show how important it is that captains should understand the theory of cyclones in both hemispheres, we shall here relate the manner in which Captain J. V. Hall escaped from a typhoon of the China seas. About noon, when three days out from Macao, Captain Hall saw "a most wild and uncommon-looking halo round the sun." On the afternoon of the next day, the barometer had commenced to fall rapidly; and though, as yet, the weather was fine, orders were at once given to prepare for a heavy gale. Toward evening, a bank of cloud was seen in the southeast, but when night closed, the weather was still calm and the water smooth, though the sky looked wild and a scud was coming on from the northeast. "I was much interested," says Captain Hall, "in watching for the commencement of the gale, which I now felt sure was coming. That bank to the southeast was the meteor (cyclone) approaching us, the northeast scud the outer northwest portion of it; and when at night a strong gale came on about north, or north-north-west, I felt certain we were on its western and south-western verge. It rapidly increased in violence; but I was pleased to see the wind veering to the north-west, as it convinced me that I had put the ship on the right track, namely, on the starboard tack, standing, of course, to the southwest. From ten A. M. to three P. M. it blew with great violence, but the ship being well prepared rode comparatively easy. The barometer was now very low, the centre of the storm passing to the northward of us, to which we might have been very near had we in the first place put the ship on the larboard tack."

But the most remarkable point of Captain Hall's account remains to be mentioned. He had gone out of his course to avoid the storm, but when the wind fell to a moderate gale he thought it a pity to lie so far from his proper course, and made sail to the northwest. "In less than two hours the barometer again began to fall and the storm to rage in heavy gusts." He bore again to the southeast, and the weather rapidly improved. There can be little doubt that but for Captain Hall's knowledge of the law of cyclones, his ship and crew would have been placed in serious jeopardy, since in the heart of a Chinese typhoon a ship has been known to be thrown on her beams-ends when not showing a yard of canvas.

If we consider the regions in which cyclones appear, the paths they follow, and the direction in which they whirl, we shall be able to form an opinion as to their origin. In the open Pacific Ocean (as its name, indeed, implies) storms are uncommon; they are infrequent also in the South Atlantic and South Indian Oceans. Around Cape Horn and the Cape of Good Hope, heavy storms prevail, but they are not cyclonic, nor are they equal in fury and frequency, Maury tells us, to the true tornado. Along the equator, and for several degrees on either side of it, cyclones are also unknown. If we turn to a map in which ocean-currents are laid down, we shall see that in every "cyclone region" there is a strongly-marked current, and that each current follows closely the track which we have denominated the storm-⊏. In the North Atlantic we have the great Gulf Stream, which sweeps from equatorial regions into the Gulf of Mexico, and thence across the Atlantic to the shores of Western Europe. In the South Indian Ocean there is the "south equatorial current" which sweeps past Mauritius and Bourbon, and thence returns toward the east. In the Chinese Sea, there is the north equatorial current, which sweeps round the East-Indian Archipelago, and then merges into the Japanese current. There is also the current in the Bay of Bengal, flowing through the region in which, as we have seen, cyclones are commonly met with. There are other sea-currents besides these which yet breed no cyclones. But we may notice two peculiarities in the currents we have named. They all flow from equatorial to temperate regions, and, secondly, they are all "horseshoe currents." So far as we are aware, there is but one other current which presents both these pecularities, namely—the great Australian current between New Zealand and the eastern shores of Australia. We have not yet met with any record of cyclones occurring over the Australian current, but heavy storms are known to prevail in that region, and we believe that when these storms have been studied as closely as the storms in better-known regions, they will be found to present the true cyclonic character.

Now, if we inquire why an ocean-current travelling from the equator should be a "storm-breeder," we shall find a ready answer. Such a current, carrying the warmth of intertropical regions to the temperate zones, produces in the first place, by the mere difference of temperature, important atmospheric disturbances. The difference is so great, that Franklin suggested the use of the thermometer in the North Atlantic Ocean as a ready means of determining the longitude, since the position of the Gulf Stream at any given season is almost constant.

Bnt the warmth of the stream itself is not the only cause of atmospheric disturbance. Over the warm water vapor is continually rising; and, as it rises, is continually condensed (like the steam from a locomotive) by the colder air round. "An observer on the moon," says Captain Maury, "would, on a winter's day, be able to trace out by the mist in the air, the path of the Gulf Stream through the sea." But what must happen when vapor is condensed? We know that to turn water into vapor is a process requiring—that is, *using up*—a large amount of heat; and, conversely, the return of vapor to the state of water *sets free* an equivalent quantity of heat. The amount of heat thus set free from the Gulf Stream is thousands of times greater than that which would be generated by the whole coal-supply annually raised in Great Britain. Here, then, we have an efficient cause for the wildest hurricanes. For along the whole of the Gulf Stream, from Bemini to the Grand Banks, there is a channel of heated—that is, *rarefied air*. Into this channel the denser atmosphere on both sides is continually pouring, with greater or less strength. When a storm begins in the Atlantic, it always makes for this channel, "and, reaching it, turns and follows it in its course, sometimes entirely across the Atlantic." "The southern points of America and Africa have won for themselves," says Maury, "the name of 'the stormy capes,' but there is not a storm-fiend in the wide ocean can out-top that which rages along the Atlantic coasts of North America. The China seas and the North Pacific may vie in the fury of their gales with this part of the Atlantic, but Cape Horn and the Cape of Good Hope cannot equal them, certainly in frequency, nor do I believe, in fury." We read of a West-Indian storm so violent, that "it forced the Gulf Stream back to its sources, and piled up the water to a height of thirty feet in the Gulf of Mexico. The ship 'Ledbury Snow' attempted to ride out the storm. When it abated, she found herself high up on the dry land, and discovered that she had let go her anchor among the tree-tops on Elliott's Key."

By a like reasoning we can account for the cyclonic storms prevailing in the North Pacific Ocean. Nor do the tornadoes which rage in parts of the United States present any serious difficulty. The region along which these storms travel is the valley of the great Mississippi. This river at certain seasons is considerably warmer than the surrounding lands. From its surface, also, aqueous vapor is continually being raised. When the surrounding air is colder, this vapor is presently condensed, generating in the change a vast amount of heat. We have thus a channel of rarefied air over the Mississippi Valley, and this channel becomes a storm-track, like the corresponding channels over the warm ocean-currents. The extreme violence of land-storms is probably due to the narrowness of the track within which they are compelled to travel. For it has been noticed that the fury of a sea-cyclone increases as the range of the "whirl" diminishes, and *vice versa*.

There seems, however, no special reason why cyclones should follow the storm-⊏ in one direction rather than in the other. We must, to understand this, recall the fact that under the torrid zones the conditions necessary for the generation of storms prevail far more intensely than in temperate regions. Thus the probability is far greater that cyclones should be generated at the tropical than at the temperate end of the storm-⊏. Still it is worthy of notice, that in the land-locked North Pacific Ocean, true typhoons *have* been known to follow the storm-track in a direction contrary to that commonly noticed.

The direction in which a true tornado *whirls* is *invariably* that we have mentioned. The explanation of this peculiarity would occupy more space than we can here afford. Those readers who may wish to understand the origin of the law of cyclonic rotation should study Herschel's interesting work on Meteorology.

The suddenness with which a true tornado works destruction was strikingly exemplified in the wreck of the steamship "San Francisco." She was assailed by an extra tropical tornado when about 300 miles from Sandy Hook, on December 24, 1853. In a few moments she was a complete wreck! The wide range of a tornado's destructiveness is shown by this, that Colonel Reid examined one along whose track no less than 110 ships were wrecked, crippled, or dismasted.

(From *Temple Bar*, December, 1867.)

---

## *VESUVIUS.*

The eruption in progress, as we write, from Mount Vesuvius, and the numerous and violent eruptions from this mountain during the last two centuries, seem to afford an answer to those who think there are traces of a gradually diminishing activity in the earth's internal forces. That such a diminution is taking place, we may admit; but that its rate of progress is perceptible —that we can point to a time within the historical epoch, nay even within the limits of geological evidence, at which the earth's internal forces were *cer-*

*tainly* more active than they are at the present time—may, we think, be denied absolutely.

When the science of geology was but young, and its professors sought to compress within a few years (at the outside) a series of events which (we now know) must have occupied many centuries, there was room, indeed, for the supposition that modern volcanic eruptions, as compared with ancient outbursts, are but as the efforts of children compared with the work of giants. And accordingly, we find a distinguished French geologist writing, even so late as 1829, that in ancient times "tous les phénomènes géologiques se passaient dans des dimensions *centuples* de celles qu'ils présentent aujourd'hui." But now we have such certain evidence of the enormous length of the intervals within which volcanic regions assumed their present appearance—we have such satisfactory means of determining which of the events occurring within those intervals were or were not contemporary—that we are safe from the error of assuming that Nature at a single effort fashioned widely-extended districts just as we now see them. And accordingly, we have the evidence of one of the most distinguished of living geologists, that there is no volcanic mass "of ancient date, distinctly referable to a single eruption, which can even *rival* in volume the matter poured out from Skaptâr Jokul in 1783."

In the volcanic region of which Vesuvius or Somma

is the principal vent, we have a remarkable instance of the deceptive nature of that state of rest into which some of the principal volcanoes frequently fall for many centuries together. For how many centuries before the Christian era Vesuvius had been at rest, is not known; but this is certain, that, from the landing of the first Greek colony in Southern Italy, Vesuvius gave no signs of internal activity. It was recognized by Strabo as a volcanic mountain, but Pliny did not include it in the list of active volcanoes. In those days, the mountain presented a very different appearance from that which it now exhibits. In place of the two peaks now seen, there was a single, somewhat flattish, summit, on which a slight depression marked the place of an ancient crater. The fertile slopes of the mountain were covered with well-cultivated fields, and the thriving cities Herculaneum, Pompeii, and Stabiæ, stood near the base of the sleeping mountain. So little did any thought of danger suggest itself in those times, that the bands of slaves, murderers, and pirates which flocked to the standard of Spartacus found a refuge, to the number of many thousands, within the very crater itself.

But though Vesuvius was at rest, the region of which Vesuvius is the main vent was far from being so. The island of Pithecusa (the modern Ischia) was skaken by frequent and terrible convulsions. It is even related that Prochyta (the modern Procida) was rent from Pithecusa in the course of a tremendous upheaval, though Pliny derives the name Prochyta (or "poured forth") from the supposed fact of this island having been poured forth by an eruption from Ischia. Far more probably, Prochyta was formed independently by submarine eruptions, as the volcanic islands near Santorin have been produced in more recent times.

So fierce were the eruptions from Pithecusa, that several Greek colonies which attempted to settle on this island were compelled to leave it. About 380 years before the Christian era, colonists under King Hiero of Syracuse, who had built a fortress on Pithecusa, were driven away by an eruption. Nor were eruptions the sole cause of danger. Poisonous vapors, such as are emitted by volcanic craters after eruption, appear to have exhaled, at times, from extensive tracts on Pithecusa, and thus to have rendered the island uninhabitable.

Still nearer to Vesuvius lay the celebrated Lake Avernus. The name Avernus is said to be a corruption of the Greek word *Aornos*, signifying "without birds," the poisonous exhalations from the waters of the lake destroying all birds which attempted to fly over its surface. Doubt has been thrown on the destructive properties assigned by the ancients to the vapors ascending from Avernus. The lake is now a healthy and agreeable neighborhood, frequented, says Humboldt, by many kinds of birds, which suffer no injury whatever, even when they skim the very surface of the water. Yet there can be little doubt that Avernus hides the outlet of an extinct volcano; and long after this volcano had become inactive, the lake which concealed its site "may have deserved the appellation of 'atri janua Ditis,' emitting, perhaps, gases as destructive of animal life as those suffocating vapors given out by Lake Quilotoa, in Quito, in 1797, by which whole herds of cattle were killed on its shores, or as those deleterious emanations which annihilated all the cattle in the island of Lancerote, one of the Canaries, in 1730."

While Ischia was in full activity, not only was Vesuvius quiescent, but even Etna seemed to be gradually expiring, so that Seneca ranks this volcano among the number of nearly-extinguished craters. At a later epoch, Ælian asserted that the mountain itself was sinking, so that seamen lost sight of the summit at a less distance across the seas than of old. Yet within the last two hundred years there have been eruptions from Etna rivalling, if not surpassing, in intensity the convulsions recorded by ancient historians.

We shall not here attempt to show that Vesuvius and Etna belong to the same volcanic sytem, though there is reason not only for supposing this to be the case, but for the belief that all the subterranean regions whose effects have been shown from time to time over the district extending from the Canaries and the Azores, across the whole of the Mediterranean, and into Syria itself, belong to but one great centre of internal action. But it is quite certain that Ischia and Vesuvius are outlets from a single source.

While Vesuvius was dormant, resigning for a while its pretensions to be the principal vent of the great Neapolitan volcanic system, Ischia, we have seen, was rent by frequent convulsions. But the time was approaching when Vesuvius was to resume its natural functions, and with all the more energy that they had been for a while suspended.

In the year 63 (after Christ) there occurred a violent convulsion of the earth around Vesuvius, during which much injury was done to neighboring cities, and many lives were lost. From this period shocks of earthquake were felt from time to time for sixteen years. These grew gradually more and more violent, until it began to be evident that the volcanic fires were about to return to their main vent. The obstruction which had so long impeded the exit of the confined matter was not, however, readily removed, and it was only in August of the year 79, after numerous and violent internal throes, that the superincumbent mass was at length hurled forth. Rocks and cinders, lava, sand, and scoriæ, were propelled from the crater, and spread many miles on every side of Vesuvius.

We have an interesting account of the great

eruption which followed in a letter from the younger Pliny to the younger Tacitus. The latter had asked for an account of the death of the elder Pliny, who lost his life in his eagerness to obtain a near view of the dreadful phenomenon. "He was at that time," says his nephew, "with the fleet under his command at Misenum. On August 24th, about one in the afternoon, my mother desired him to observe a cloud of very extraordinary size and shape. He had just returned from taking the benefit of the sun, and, after bathing himself in cold water, and taking a slight repast, had retired to his study. He arose at once, and went out upon a height whence he might more distinctly view this strange phenomenon. It was not at this distance discernible from what mountain the cloud issued, but it was found afterward that it came from Vesuvius. I cannot give a more exact description of its figure than by comparing it to that of a pine-tree, for it shot up to a great height in the form of a trunk, which extended itself at the top into a sort of branches; occasioned, I suppose, either by a sudden gust of air which impelled it, whose force decreased as it advanced upward, or else the cloud itself, being pressed back by its own weight, expanded in this manner. The cloud appeared sometimes bright, at others dark and spotted, as it was more or less impregnated with earth and cinders."

These extraordinary appearances attracted the curiosity of the elder Pliny. He ordered a small vessel to be prepared, and started to seek a nearer view of the burning mountain. His nephew declined to accompany him, being engaged with his studies. As Pliny left the house, he received a note from a lady whose house, being at the foot of Vesuvius, was in imminent danger of destruction. He set out, accordingly, with the design of rendering her assistance, and also of assisting others, "for the villas stood extremely thick upon that lovely coast." He ordered the galleys to be put to sea, and steered directly to the point of danger, so cool in the midst of the turmoil around "as to be able to make and dictate observations upon the motions and figures of that dreadful scene." As he approached Vesuvius, cinders, pumice-stones, and black fragments of burning rock, fell on and around the ships. "They were in danger, too, of running aground, owing to the sudden retreat of the sea; vast fragments, also, rolled down from the mountain and obstructed all the shore." The pilot advising retreat, Pliny made the noble answer, "Fortune befriends the brave," and bade him press onward to Stabiæ. Here he found his friend Pomponianus in great consternation, already prepared for embarking, and waiting only for a change in the wind. Exhorting Pomponianus to be of good courage, Pliny quietly ordered baths to be prepared; and "having bathed, sat down to supper with great cheerfulness, or at least (which is equally heroic) with all the appearance of it." Assuring his friends that the flames which appeared in several places were merely burning villages, Pliny presently retired to rest, and "being pretty fat," says his nephew, "and breathing hard, those who attended without actually heard him snore." But it became necessary to awaken him, for the court which led to his room was now almost filled with stones and ashes. He got up and joined the rest of the company, who were consulting on the propriety of leaving the house, now shaken from side to side by frequent concussions. They decided on seeking the fields for safety; and fastening pillows on their heads, to protect them from falling stones, they advanced in the midst of an obscurity greater than that of the darkest night—though beyond the limits of the great cloud it was already broad day. When they reached the shore, they found the waves running too high to suffer them safely to venture to put out to sea. Pliny, "having drunk a draught or two of cold water, lay down on a cloth that was spread out for him; but at this moment the flames and sulphurous vapors dispersed the rest of the company and obliged him to rise. Assisted by two of his servants, he got upon his feet, but instantly fell down dead; suffocated, I suppose," says his nephew, "by some gross and noxious vapor, for he always had weak lungs and suffered from a difficulty of breathing." His body was not found until the third day after his death, when for the first time it was light enough to search for him. He was found as he had fallen, "and looking more like a man asleep than dead."

But even at Misenum there was danger, though Vesuvius is distant no less than fourteen miles. The earth was shaken with repeated and violent shocks, "insomuch," says the younger Pliny, "that they threatened our complete destruction." When morning came, the light was faint and glimmering; the buildings around seemed tottering to their fall, and, standing on the open ground, the chariots which Pliny had ordered were so agitated backward and forward that it was impossible to keep them steady, even by supporting them with large stones. The sea was rolled back upon itself, and many marine animals were left dry upon the shore. On the side of Vesuvius, a black and ominous cloud, bursting with sulphurous vapors, darting out long trains of fire, resembling flashes of lightning, but much larger. Presently the great cloud spread over Misenum and the island of Capreæ. Ashes fell around the fugitives. On every side "nothing was to be heard but the shrieks of women and children, and the cries of men: some were calling for their children, others for their parents, others for their husbands, and only distinguishing each other by their voices: one was lamenting his own fate, another that of his family; some wished to die, that they might escape the dreadful fear of death;

but the greater part imagined that the last and eternal night was come, which was to destroy the gods and the world together." At length a light appeared, which was not, however, the day, but the forerunner of an outburst of flames. These presently disappeared, and again the thick darkness spread over the scene. Ashes fell heavily upon the fugitives, so that they were in danger of being crushed and buried in the thick layer rapidly covering the whole country. Many hours passed before the dreadful darkness began slowly to be dissipated. When at length day returned, and the sun was seen faintly shining through the overhanging canopy of ashes, "every object seemed changed, being covered over with white ashes as with a deep snow."

It is most remarkable that Pliny makes no mention in his letter of the destruction of the two populous and important cities, Pompeii and Herculaneum. We have seen that at Stabiæ a shower of ashes fell so heavily that several days before the end of the eruption the court leading to the elder Pliny's room was beginning to be filled up; and when the eruption ceased, Stabiæ was completely overwhelmed. Far more sudden, however, was the destruction of Pompeii and Herculaneum.

It would seem that the two cities were first shaken violently by the throes of the disturbed mountain. The signs of such a catastrophe have been very commonly assigned to the earthquake which happened in 63, but it seems far more likely that most of them belong to the days immediately preceding the great outburst in 79. "In Pompeii," says Sir Charles Lyell, "both public and private buildings bear testimony to the catastrophe. The walls are rent, and in many places traversed by fissures still open." It is probable that the inhabitants were driven by these anticipatory throes to fly from the doomed towns. For though Dion Cassius relates that "two entire cities, Herculaneum and Pompeii, were buried under showers of ashes, while all the people were sitting in the theatre," yet "the examination of the two cities enables us to prove," says Sir Charles, "that none of the people were destroyed in the theatre, and, indeed, that there were very few of the inhabitants who did not escape from both cities. Yet," he adds, "some lives were lost, and there was ample foundation for the tale in all its most essential particulars."

We may note here, in passing, that the account of the eruption given by Dion Cassius, who wrote a century and a half after the catastrophe, is sufficient to prove how terrible an impression had been made upon the inhabitants of Campania, from whose descendants he in all probability obtained the materials of his narrative. He writes that, "during the eruption, a multitude of men of superhuman stature, resembling giants, appeared, sometimes on the mountain, and sometimes in the environs; that stones and smoke were thrown out, the sun was hidden, and then the giants seemed to rise again, while the sounds of trumpets were heard"—with much other matter of a similar sort.

In the great eruption of 79, Vesuvius poured forth lapilli, sand, cinders, and fragments of old lava, but no new lava flowed from the crater. Nor does it appear that any lava-stream was ejected during the six eruptions which took place during the following ten centuries. In the year 1036, for the first time, Vesuvius was observed to pour forth a stream of molten lava. Thirteen years later, another eruption took place; then 90 years passed without disturbance, and after that a long pause of 168 years. During this interval, however, the volcanic system of which Vesuvius is the main but not the only vent, had been disturbed twice. For it is related that in 1198 the Solfatara Lake crater was in eruption; and in 1302, Ischia, dormant for at least 1,400 years, showed signs of new activity. For more than a year earthquakes had convulsed this island from time to time, and at length the disturbed region was relieved by the outburst of a lava-stream from a new vent on the south-east of Ischia. The lava-stream flowed right down to the sea, a distance of two miles. For two months, this dreadful outburst continued to rage; many houses were destroyed; and although the inhabitants of Ischia were not completely expelled, as happened of old with the Greek colonists, yet a partial emigration took place.

The next eruption of Vesuvius occurred in 1306; and then three centuries and a quarter passed during which only one eruption, and that an unimportant one (in 1500), took place. "It was remarked," says Sir Charles Lyell, "that throughout this long interval of rest, Etna was in a state of unusual activity, so as to lend countenance to the idea that the great Sicilian volcano may sometimes serve as a channel of discharge to elastic fluids and lava that would otherwise rise to the vents in Campania."

Nor was the abnormal activity of Etna the only sign that the quiescence of Vesuvius was not to be looked upon as any evidence of declining energy in the volcanic system. In 1538 a new mountain was suddenly thrown up in the Phlegræan Fields—a district including within its bounds Pozzuoli, Lake Avernus, and the Solfatara. The new mountain was thrown up near the shores of the Bay of Baiæ. It is 440 feet above the level of the bay, and its base is about a mile and a half in circumference. The depth of the crater is 421 feet, so that its bottom is only six yards above the level of the bay. The spot on which the mountain was thrown up was formerly occupied by the Lucrine Lake; but the outburst filled up the greater part of the lake, leaving only a small and shallow pool.

The accounts which have reached us of the formation of this new mountain are not without interest. Falconi, who wrote in 1538, mentions that several earthquakes took place during the two years preceding the outburst, and above twenty shocks on the day and night before the eruption. "The eruption began on September 29, 1538. It was on a Sunday, about one o'clock in the night, when flames of fire were seen between the hot-baths and Tripergola. In a short time the fire increased to such a degree that it burst open the earth in this place, and threw up a quantity of ashes and pumice-stones, mixed with water, which covered the whole country. The next morning the poor inhabitants of Pozzuoli quitted their habitations in terror, covered with the muddy and black shower, which continued the whole day in that country—flying from death, but with death painted in their countenances. Some with their children in their arms, some with sacks full of their goods; others leading an ass, loaded with their frightened family, toward Naples. . . . . The sea had retired on the side of Baiæ, abandoning a considerable tract; and the shore appeared almost entirely dry, from the quantity of ashes and broken pumice-stones thrown up by the eruption."

Pietro Giacomo di Toledo gives us some account of the phenomena which preceded the eruption: "That plain which lies between Lake Avernus, the Monte Barbaro, and the sea, was raised a little, and many cracks were made in it, from some of which water issued; at the same time the sea immediately adjoining the plain dried up about two hundred paces, so that the fish were left on the sand, a prey to the inhabitants of Pozzuoli. At last, on September 29th, about two o'clock in the night, the earth opened near the lake, and discovered a horrid mouth, from which were furiously vomited smoke, fire, stones, and mud composed of ashes, making at the time of the opening a noise like the loudest thunder. The stones which followed were by the flames converted to pumice, and some of these were *larger than an ox.* The stones went about as high as a cross-bow will carry, and then fell down, sometimes on the edge, and sometimes into the mouth itself. The mud was of the color of ashes, and at first very liquid, then by degrees less so; and in such quantities that in less than twelve hours, with the help of the above-mentioned stones, a mountain was raised of 1,000 paces in height. Not only Pozzuoli and the neighboring country were full of this mud, but the city of Naples also; so that many of its palaces were defaced by it. This eruption lasted two nights and two days without intermission, though not always with the same force; the third day the eruption ceased, and I went up with many people to the top of the new hill, and saw down into its mouth, which was a round cavity about a quarter of a mile in circumference, in the middle of which the stones which had fallen were boiling up just as a caldron of water boils on the fire. The fourth day it began to throw up again, and the seventh day much more, but still with less violence than the first night. At this time many persons who were on the hill were knocked down by the stones and killed, or smothered with the smoke."

And now, for nearly a century, the whole district continued in repose. Nearly five centuries had passed since there had been any violent eruption of Vesuvius itself; and the crater seemed gradually assuming the condition of an extinct volcano. The interior of the crater is described by Bracini, who visited Vesuvius shortly before the eruption of 1631, in terms that would have fairly represented its condition before the eruption of 79: "The crater was five miles in circumference, and about a thousand paces deep; its sides were covered with brushwood, and at the bottom there was a plain on which cattle grazed. In the woody parts, wild boars frequently harbored. In one part of the plain, covered with ashes, were three small pools, one filled with hot and bitter water, another salter than the sea, and a third hot, but tasteless." But in December, 1631, the mountain blew away the covering of rock and cinders which supported these woods and pastures. Seven streams of lava poured from the crater, causing a fearful destruction of life and property. Resina, built over the site of Herculaneum, was entirely consumed by a raging lava-stream. Heavy showers of rain, generated by the steam evolved during the eruption, caused in their turn an amount of destruction scarcely less important than that resulting from the lava-streams. For, falling upon the cone, and sweeping thence large masses of ashes and volcanic dust, these showers produced destructive streams of mud, consistent enough to merit the name of "aqueous lava" commonly assigned to it.

An interval of thirty-five years passed before the next eruption. But, since 1666, there has been a continual series of eruptions, so that the mountain has scarcely ever been at rest for more than ten years together. Occasionally there have been two eruptions within a few months; and it is well worthy of remark that, during the three centuries which have elapsed since the formation of Monte Nuovo, there has been no volcanic disturbance in any part of the Neapolitan volcanic district save in Vesuvius alone. Of old, as Brieslak well remarks, there had been irregular disturbances in some part of the Bay of Naples once in every two hundred years; the eruption of Solfatara in the twelfth century, that of Ischia in the fourteenth, and that of Monte Nuovo in the sixteenth; but "the eighteenth has formed an exception to the rule." It seems clear that the constant series of eruptions from Vesuvius during the past two hundred years has sufficed to relieve the volcanic district of which Vesuvius is the principal vent.

Of the eruptions which have disturbed Vesuvius during the last two centuries, those of 1779, 1793, and 1822, are in some respects the most remarkable.

Sir William Hamilton has given a very interesting account of the eruption of 1779. Passing over those points in which this eruption resembled others, we may note its more remarkable features. Sir William Hamilton says, that in this eruption molten lava was thrown up in magnificent jets to the height of at least 10,000 feet. Masses of stones and scoriæ were to be seen propelled along by these lava-jets. Vesuvius seemed to be surmounted by an enormous column of fire. Some of the jets were directed by the wind toward Ottajano; others fell on the cone of Vesuvius, on the outer circular mountain Somma, and on the valley between. Falling, still red hot and liquid, they covered a district more than two miles and a half wide with a mass of fire. The whole space above this district, to the height of 10,000 feet, was filled also with the falling and rising lava-streams; so that there was continually present a body of fire covering the extensive space we have mentioned, and extending nearly two miles high. The heat of this enormous fire-column was distinctly perceptible at a distance of at least six miles on every side.

The eruption of 1793 presented a different aspect. Dr. Clarke tells us that millions of red-hot stones were propelled into the air to at least half the height of the cone itself; then turning, they fell all around in noble curves. They covered nearly half the cone of Vesuvius with fire. Huge masses of white smoke were vomited forth by the disturbed mountain, and formed themselves, at a height of many thousands of feet above the crater, into a huge, ever-moving canopy, through which, from time to time, were hurled pitch-black jets of volcanic dust, and dense vapors, mixed with cascades of red-hot rocks and scoriæ. The rain which fell from the cloud-canopy was scalding hot.

Dr. Clarke was able to compare the different appearances presented by the lava when it burst from the very mouth of the crater, and lower down when it had approached the plain. As it rushed forth from its imprisonment, it streamed a liquid, white, and brilliantly pure river, which burned for itself a smooth channel through a great arched chasm in the side of the mountain. It flowed with the clearness of "honey in regular channels, cut finer than art can imitate, and glowing with all the splendor of the sun. Sir William Hamilton had conceived," adds Dr. Clarke, "that stones thrown upon a current of lava would produce no impression. I was soon convinced of the contrary. Light bodies, indeed, of five, ten, and fifteen pounds' weight, made little or no impression, even at the source; but bodies of sixty, seventy, and eighty pounds were seen to form a kind of bed on the surface of the lava, and floated away with it. A stone of three hundred-weight, that had been thrown out by the crater, lay near the source of the current of lava. I raised it up on one end, and then let it fall in upon the liquid lava, when it gradually sank beneath the surface and disappeared. If I wished to describe the manner in which it acted upon the lava, I should say that it was like a loaf of bread thrown into a bowl of very thick honey, which gradually involves itself in the heavy liquid, and then slowly sinks to the bottom."

But, as the lava flowed down the mountain-slopes, it lost its brilliant whiteness; a crust began to form upon the surface of the still molten lava, and this crust broke into innumerable fragments of porous matter, called scoriæ. Underneath this crust—across which Dr. Clarke and his companions were able to pass without other injury than the singeing of their boots—the liquid lava still continued to force its way onward and downward past all obstacles. On its arrival at the bottom of the mountain, says Dr. Clarke, "the whole current," encumbered with huge masses of scoriæ, "resembled nothing so much as a heap of unconnected cinders from an iron-foundery," "rolling slowly along," he says in another place, "and falling with a rattling noise over one another."

After the eruption described by Dr. Clarke, the great crater gradually filled up. Lava boiled up from below, and small craters, which formed themselves over the bottom and sides of the great one, poured forth lava loaded with scoriæ. Thus, up to October, 1822, there was to be seen, in place of a regular crateriform opening, a rough and uneven surface, scored by huge fissures, whence vapor was continually being poured, so as to form clouds above the hideous heap of ruins. But the great eruption of 1822 not only flung forth all the mass which had accumulated within the crater, but wholly changed the appearance of the cone. An immense abysm was formed, three-quarters of a mile across, and extending 2,000 feet downward into the very heart of Vesuvius. Had the lips of the crater remained unchanged, indeed, the depth of this great gulf would have been far greater. But so terrific was the force of the explosion that the whole of the upper part of the cone was carried clean away, and the mountain reduced in height by nearly a full fifth of its original dimensions. From the time of its formation the chasm gradually filled up; so that, when Mr. Scrope saw it soon after the eruption, its depth was reduced by more than 1,000 feet.

Of late, Vesuvius has been as busy as ever. In 1833 and 1834 there were eruptions; and it is but twelve years since a great outburst took place. Then, fcr three weeks together, lava streamed down the mountain-slopes. A river of molten lava swept away the village of Cercolo, and ran nearly to the sea at Ponte Maddaloni. There were then formed ten small craters within the great one. But these have now united, and pressure from beneath has formed a vast cone where they had been. The cone has risen above the rim of the crater, and as we write torrents of lava are being poured forth. At first the lava formed a lake of fire, but the seething mass found an outlet, and poured in a wide stream toward Ottajano. Masses of red-hot stone and rock are hurled forth, and a vast canopy of white vapor hangs over Vesuvius, forming at night, when illuminated by the raging mass below, a glory of resplendent flame around the summit of the mountain.

It may seem strange that the neighborhood of so dangerous a mountain should be inhabited by races free to choose more peaceful districts. Yet, though Herculaneum, Pompeii, and Stabiæ, lie buried beneath the lava and ashes thrown forth by Vesuvius, Portici and Resina, Torre del Greco and Torre dell' Annunziata have taken their place; and a large population, cheerful and prosperous, flourish around the disturbed mountain, and over the district of which it is the somewhat untrustworthy safety-valve.

It has, indeed, been well pointed out by Sir Charles Lyell that "the general tendency of subterranean movements, when their effects are considered for a sufficient lapse of ages, is eminently beneficial, and that they constitute an essential part of that mechanism by which the integrity of the habitable surface is preserved. Why the working of this same machinery should be attended with so much evil, is a mystery far beyond the reach of our philosophy, and must probably remain so until we are permitted to investigate, not our planet alone and its inhabitants, but other parts of the moral and material universe with which they may be connected. Could our survey embrace other worlds, and the events, not of a few centuries only, but of periods as indefinite as those with which geology renders us familiar, some apparent contradictions might be reconciled, and some difficulties would doubtless be cleared up. But even then, as our capacities are finite, while the scheme of the universe may be infinite, both in time and space, it is presumptuous to suppose that all source of doubt and perplexity would ever be removed. On the contrary, they might, perhaps, go on augmenting in number, although our confidence in the wisdom of the plan of Nature should increase at the same time; for it has been justly said" (by Sir Humphry Davy) "that the greater the circle of light, the greater the boundary of darkness by which it is surrounded."

(From the *Cornhill Magazine*, March, 1868.)

---

## *THE EARTHQUAKE IN PERU.*

THE intelligence published last Saturday is sufficient to prove that the great earthquake which has devastated Peru fully equalled, if it did not surpass, the most terrible catastrophes which have ever befallen that country. It presents, too, all the features which have hitherto characterized earthquakes in this neighborhood. These are well worthy of careful study, and appear to have an important bearing on the modern theory of earthquakes.

It has been commonly held that the seat of disturbance in the earthquakes which have shaken the country west of the Andes has lain always at some point or other beneath that range of mountains. The fact that several large volcanoes are found in the Cordilleras has seemed confirmatory of this view. The accounts we have also of the great earthquake at Riobamba in 1797, seem only explicable by supposing that the seat of disturbance lay almost immediately beneath that city. The inhabitants were flung vertically upward into the air, and to such a height that Humboldt found the skeletons of many of them on the summit of the hill La Culca, on the farther side of the small river on which Riobamba is built. The ruins of many houses were also flung to the same spot. Here, therefore, was evidence of that vertical (or, as Humboldt expresses it, explosive) force which is only to be looked for immediately above the centre of concussion.

Yet the consideration of the evidence afforded by the news we have just published, seems at first sight somewhat opposed to this view, and to point rather to a seat of disturbance lying considerably to the west of the Peruvian shores. "At Chala," says our informant, "the sea receded, and a wave rose fifty feet, and returned, spreading into the town a distance of about a thousand feet. Three successive times every thing within range was swept away, followed by twelve shocks of earthquake, lasting from three seconds to two minutes." The arrival of great sea-waves before the land-shocks were felt seems decisively to indicate that the seat of disturbance lay beneath the ocean and not beneath the land. We are disposed to believe, however, that in the confusion of mind naturally resulting from the occurrence of so terrible a catastrophe, the sequence of events may not have been very closely attended to, for in other places the arrival of the great sea-wave is distinctly described as following the occurrence of the earth-shock. At Arica, for example, a considerable interval would seem to have elapsed before the terrible sea-wave, which has always characterized Peruvian earthquakes, poured in upon the town. The agent of the Pacific Steam Navigation Company, whose house had been destroyed by the earth-shock, saw the great sea-wave while he was flying toward the hills. He writes: "While passing toward the hills, with the earth shaking, a great cry went up to heaven. The sea had retired. On clearing the town, I looked back and saw that the vessels were being carried irresistibly seaward. In a few minutes the sea stopped, and then arose a mighty wave fifty feet high, and came in with a fearful rush, carrying every thing before it in terrible majesty. The whole of the shipping came back, speeding toward inevitable doom. In a few minutes all was completed—every vessel was either on shore or bottom upward." This, then, was undoubtedly the great sea-wave, as compared with the minor waves of disturbance which characterize all earthquakes near the shores of the ocean.

One remarkable feature in this terrible earthquake is the enormous range of country affected by it. From Quito southward as far as Iquique—or, in other words, for a distance considerably exceeding a full third part of the whole length of the South American Andes—the shock was felt with the most terrible distinctness. We have yet to learn how much farther to the north and south, and how far inland on the eastern slopes of the Andes, the shock was experienced. But there can be little doubt that the disturbed country was equal to at least a fourth of Europe.

The portion of the Andes thus disturbed seems to be distinct from the part to which the great Chilian earthquakes belong. The difference in character between the Peruvian and Chilian earthquakes is a singular and interesting phenomenon. The difference corresponds to a feature long since pointed out by Sir Charles Lyell—the alternation, on a grand scale, of districts of active with those of extinct volcanoes. It is said that in Chili a year scarcely ever passes without shocks of earthquake being felt; in certain regions, not even a month. A similar persistence of earthquake disturbance characterizes Peru. Yet, although both districts are shaken in this manner, there seems to be a distinct evidence of alternating disturbance as respects the occurrence of great earthquakes. Thus, in 1797, took place the terrible earthquake of Riobamba. Then, thirty years later, a series of great earthquakes shook Chili, permanently elevating the whole line of coast to the height of several feet. Now, again, after another interval of about thirty years, the Andes are disturbed by a great earthquake, and this time it is the Peruvian Andes which experience the shock. Between Chili and Peru there is a space upward of five hundred miles long, in which no volcanic action has been observed. Singularly enough, this very portion of the Andes, to which one would imagine the Peruvians and Chilians would fly as to a region of safety, is the part most thinly inhabited, insomuch that, as Von Buch observes, it is in some places entirely deserted.

Near Quito the trembling of the earth is almost incessant, according to M. Boussingault. He considers that the frequency of the movement is due rather to the continual falling in of masses of rock which have been fractured in recent earthquakes, than to the persistence of subterranean action. He adds that the height of several mountains in the Andes has diminished in modern times. He refers, doubtless, to the Peruvian and Colombian Andes, and not to the Chilian. In the latter portion of the range there must be a continual increase of height, since each earthquake in Chili has produced a perceptible recession of the sea. Darwin, indeed, relates that near Valparaiso he saw beds of sea-shells belonging to recent species at a height of about a quarter of a mile abovo the present sea-level; and he concluded that the land had been raised to this height by a series of such small elevations as were observed to have taken place during the earthquakes of 1822, 1835, and 1837. That a contrary process should be going on in Peru, confirms the idea that a sort of undulatory or balancing motion is taking place—one long stretch of the Cordilleras rising while another is sinking. A tradition prevails among the Indians of Lican that the mountain called L'Altar, or Cassac Urcu—which means "the chief"—was once the highest of the sub-equatorial Andes, being higher even than Chimborazo; but, adds the tradition, in the reign of Quainia Abomatha, before the discovery of America, a prodigious eruption took place which lasted no less than eight years, and brought down the summit of the mountain. M. Boussingault states that the fragments of trachyte which once formed the summit of this celebrated mountain are now spread over the plain. At present Cotopaxi is the loftiest volcano of the Cordilleras, its height being no less than 18,858 feet. No mountain has ever been the seat of such terrible and destructive eruptions as those which have burst forth from Cotopaxi. The intensity of the heat which prevails during eruption will be readily gathered from the circumstance that in January, 1803, the enormous bed of snow which usually covers the cone of the volcano was dissolved in a single night.

It would seem that the Mexican volcanoes also belong to the same region of disturbances. Near the Isthmus of Panama the great Cordillera of the Andes lowers itself to the height of about 800 feet, and beyond begins the continuation of the volcanic chain in Central America and Mexico. Nor are the volcanoes of the West Indian or Caribbee Islands wholly disconnected with the region of disturbance in Southern America. And it is rather singular that even the earthquakes which have occurred in the valley of the Mississippi seem to be connected with the West Indian and South American volcanic region. The violent earthquakes which took place at New Madrid in 1812, occurred at exactly the same time as the earthquake of Paranas, "so that it is possible," says Sir Charles Lyell, "that these two points are part of one volcanic region."

(From the *Daily News*, September 18, 1868.)

## *THE GREATEST SEA-WAVE EVER KNOWN.*

On August 13, 1868, one of the most terrible calamities which has ever visited a people befell the unfortunate inhabitants of Peru. In that land earthquakes are nearly as common as rain-storms are with us; and shocks by which whole cities are changed into a heap of ruins are by no means infrequent. Yet even in Peru, "the land of earthquakes," as Humboldt has termed it, no such catastrophe as that of August, 1868, had occurred within the memory of man. It was not one city which was laid in ruins, but a whole empire. Those who perished were counted by tens of thousands, while the property destroyed by the earthquake was valued at millions of pounds sterling.

Although so many months have passed since this terrible calamity took place, scientific men have been busily engaged until quite recently in endeavoring to ascertain the real significance of the various events which were observed during and after the occurrence of the earthquake. The geographers of Germany have taken a special interest in interpreting the evidence afforded by this great manifestation of Nature's powers. Two papers have been written recently on the great earthquake of August 13, 1868, one by Professor von Hochstetter, the other by Herr von Tschudi, which present an interesting account of the various effects, by land and by sea, which resulted from the tremendous upheaving force to which the western flanks of the Peruvian Andes were subjected on that day. The effects on land, although surprising and terrible, yet only differ in degree from those which have been observed in other earthquakes. But the progress of the great sea-wave which was generated by the upheaval of the Peruvian shores and propagated over the whole of the Pacific Ocean differs altogether from any earthquake phenomena before observed. Other earthquakes have indeed been followed by oceanic disturbances; but these have been accompanied by terrestrial motions, so as to suggest the idea that they had been caused by the motion of the sea-bottom, or of the neighboring land. In no instance has it ever before been known that a well-marked wave of enormous proportions should have been propagated over the largest ocean-tract on our globe, by an earth-shock whose direct action was limited to a relatively small region, and that region not situated in the centre, but on one side of the wide area traversed by the wave.

We propose to give a brief sketch of the history of this enormous sea-wave. In the first place, however, it may be well to remind the reader of a few of the more prominent features of the great shock to which this wave owed its origin.

It was at Arequipa, at the foot of the lofty volcanic mountain Misti, that the most terrible effects of the great earthquake were experienced. Within historic times Misti has poured forth no lava-streams, but that the volcano is not extinct is clearly evidenced by the fact that in 1542 an enormous mass of dust and ashes was vomited forth from its crater. On August 13, 1868, Misti showed no signs of being disturbed. So far as their volcanic neighbor was concerned, the 44,000 inhabitants of Arequipa had no reason to anticipate the catastrophe which presently befell them. At five minutes past five an earthquake-shock was experienced, which, though severe, seems to have worked little mischief. Half a minute later, however, a terrible noise was heard beneath the earth; a second shock more violent than the first was felt; and then began a swaying motion, gradually increasing in intensity. In the course of the first minute this motion had become so violent that the inhabitants ran in terror out of their houses into the streets and squares. In the next two minutes the swaying movement had so increased that the more lightly-built houses were cast to the ground, and the flying people could scarcely keep their feet. "And now," says Von Tschudi, "there followed during two or three minutes a terrible scene. The swaying motion which had hitherto prevailed changed into fierce vertical upheaval. The subterranean roaring increased in the most terrifying manner: then were heard the heart-piercing shrieks of the wretched people, the bursting of walls, the crashing fall of houses and churches, while over all rolled thick clouds of a yellowish-black dust, which, had they been poured forth many minutes longer, would have suffocated thousands." Although the shocks had lasted but a few minutes, the whole town was destroyed. Not one building remained uninjured, and there were few which did not lie in shapeless heaps of ruins.

At Tacna and Arica, the earth-shock was less severe, but strange and terrible phenomena followed it. At the former place a circumstance occurred, the cause and nature of which yet remain a mystery. About three hours after the earthquake—in other words, at about eight o'clock in the evening—an intensely brilliant light made its appearance above the neighboring mountains. It lasted for fully half an hour, and has been ascribed to the eruption of some as yet unknown volcano.

At Arica the sea-wave produced even more destructive effects than had been caused by the earthquake. About twenty minutes after the first earth-shock, the sea was seen to retire, as if about to leave the shores wholly dry; but presently its waters returned with tremendous force. A mighty wave, whose length seemed immeasurable, was seen advancing like a dark wall upon the unfortunate town, a large part of which was overwhelmed by it. Two ships, the Peruvian corvette "America" and the United States "double-ender" "Wateree," were carried nearly half a mile to the north of Arica, beyond the railroad which runs to Tacna, and there left stranded high and dry. This enormous wave was considered by the English vice-consul at Arica to have been fully fifty feet in height.

At Chala, three such waves swept in after the first shocks of earthquake. They overflowed nearly the whole of the town, the sea passing more than half a mile beyond its usual limits.

At Islay and Iquique similar phenomena were manifested. At the former town the sea flowed in no less than five times, and each time with greater force. Afterward the motion gradually diminished, but even an hour and a half after the commencement of this strange disturbance, the waves still ran forty feet above the ordinary level. At Iquique, the people beheld the inrushing wave while it was still a great way off. A dark-blue mass of water, some fifty feet in height, was seen sweeping in upon the town with inconceivable rapidity. An island lying before the harbor was completely submerged by the great wave, which still came rushing on, black with the mud and slime it had swept from the sea-bottom. Those who witnessed its progress from the upper balconies of their houses, and presently saw its black mass rushing close beneath their feet, looked on their safety as a miracle. Many buildings were indeed washed away, and in the low-lying parts of the town there was a terrible loss of life. After passing far inland, the wave slowly returned seaward, and strangely enough, the sea, which elsewhere heaved and tossed for hours after the first great wave had swept over it, here came soon to rest.

At Callao a yet more singular instance was afforded of the effect which circumstances may have upon the motion of the sea after a great earthquake has disturbed it. In former earthquakes Callao has suffered terribly from the effects of the great sea-wave. In fact, on two occasions the whole town has been destroyed, and nearly all its inhabitants have been drowned, through the inrush of precisely such waves as flowed into the ports of Arica and Chala. But upon this occasion the centre of subterranean disturbance must have been so situated that either the wave was diverted from Callao, or more probably two waves reached Callao from different sources and at different times, so that the two undulations partly counteracted each other. Certain it is that, although the water retreated strangely from the coast near Callao, insomuch that a wide tract of the sea-bottom was uncovered, there was no inrushing wave comparable with those described above. The sea afterward rose and fell in an irregular manner, a circumstance confirming the supposition that the disturbance was caused by two distinct oscillations. Six hours after the occurrence of the earth-shock, the double oscillations seemed for a while to have worked themselves into unison, for at this time three considerable waves rolled in upon the town. But clearly these waves must not be compared with those which in other instances had made their appearance within half an hour of the earth-throes. There is little reason to doubt that if the separate oscillations had reënforced each other earlier, Callao would have been completely destroyed. As it was, a considerable amount of mischief was effected; but the motion of the sea presently became irregular again, and so continued until the morning of August 14th, when it began to ebb with some regularity. But during the 14th there were occasional renewals of the irregular motion, and several days elapsed before the regular ebb and flow of the sea were resumed.

Such were among the phenomena presented in the region where the earthquake itself was felt. It will be seen at once that within this region, or rather along that portion of the sea-coast which falls within the central region of disturbance, the true character of the sea-wave generated by the earthquake could not be recognized. If a rock fall from a lofty cliff into a comparatively shallow sea, the water around the place where the rock has fallen is disturbed in an irregular manner. The sea seems at one place to leap up and down; elsewhere one wave seems to beat against another, and the sharpest eye can detect no law in the motion of the seething waters. But presently, outside the scene of disturbance, a circular wave is seen to form, and if the motion of this wave be watched, it is seen to present the most striking contrast with the turmoil and confusion at its centre. It sweeps onward and outward in a regular undulation. Gradually it loses its circular figure (unless the sea-bottom happens to be unusually level), showing that although its motion is everywhere regular, it is not everywhere equally swift. A wave of this sort, though incomparably vaster, swept swiftly away on every side from the scene of the great earthquake near the Peruvian Andes. It has been calculated that the width of this wave varied from one million to five million feet, or roughly from 200 to 1,000 miles, while, when in mid-Pacific, the length of the wave, measured along its summit in a widely-curved path from one side to another of the great ocean, cannot have been less than 8,000 miles.

We cannot tell how deep-seated was the centre of subterranean action; but there can be no doubt it was very deep indeed, because otherwise the shock felt in towns separated from each other by hundreds of miles could not have been so nearly contemporaneous. Therefore the portion of the earth's crust upheaved must have been enormous, for the length of the region where the direct effects of the earthquake were perceived is estimated by Professor von Hochstetter at no less than 240 miles. The breadth of the region is unknown, because the slope of the Andes on one side and the ocean on the other concealed the motion of the earth's crust.

The great ocean-wave swept, as we have said, in all directions around the scene of the earth-throe. Over a large part of its course its passage was unnoted, because in the open sea the effects even of so vast an undulation could not be perceived. A ship would slowly rise as the crest of the great wave passed under her, and then as slowly sink again. This may seem strange, at first sight, when it is remembered that in reality the great sea-wave we are considering swept at the rate of three or four hundred sea-miles an hour over the larger part of the Pacific. But when the true character of ocean-waves is understood, when it is remembered that there is no transference of the water itself at this enormous rate, but simply a transmission of motion (precisely as when in a high wind waves sweep rapidly over a cornfield, while yet each corn-stalk remains fixed in the ground), it will be seen that the effects of the great sea-wave could only be perceived near the shore. Even there, as we shall presently see, there was much to convey the impression that the land itself was rising and falling rather than that the deep was moved. But among the hundreds of ships, which were sailing upon the Pacific when its length and breadth were traversed by the great sea-wave, there was not one in which any unusual motion was perceived.

In somewhat less than three hours after the occurrence of the earthquake, the ocean-wave inundated the port of Coquimbo, on the Chilian seaboard, some 800 miles from Arica. An hour or so later it had reached Constitucion, 450 miles farther south; and here for some three hours the sea rose and fell with strange violence. Farther south, along the shore of Chili, even to the island of Chiloe, the shore-wave travelled, though with continually diminishing force, owing, doubtless, to the resistance which the irregularities of the shore opposed to its progress.

The northerly shore-wave seems to have been more considerable; and a moment's study of a chart of the two Americas will show that this circumstance is highly significant. When we remember that the principal effects of the land-shock were experienced within that angle which the Peruvian Andes form with the long north-and-south line of the Chilian and Bolivian Andes, we see at once that had the centre of the subterranean action been near the scene where the most destructive effects were perceived, no sea-wave, or but a small one, could have been sent toward the shores of North America. The projecting shores of northern Peru and Ecuador could not have failed to divert the sea-wave toward the west; and though a reflected wave might have reached California, it would only have been after a considerable interval of time, and with dimensions much less than those of the sea-wave which travelled southward. When we see that, on the contrary, a wave of even greater proportions travelled toward the shores of North America, we seem forced to the conclusion that the centre of the subterranean action must have been so far to the west that the sea-wave generated by it had a free course to the shores of California.

Be this as it may, there can be no doubt that the wave which swept the shores of Southern California, rising upward of sixty feet above the ordinary sea-level, was absolutely the most imposing of all the indirect effects of the great earthquake. When we consider that even in San Pedro Bay, fully five thousand miles from the centre of disturbance, a wave twice the height of an ordinary house rolled in with unspeakable violence only a few hours after the occurrence of the earth-throe, we are most strikingly impressed with the tremendous energy of the earth's movement.

Turning to the open ocean, let us track the great wave on its course past the multitudinous islands which dot the surface of the great Pacific.

The inhabitants of the Sandwich Islands, which lie about 6,300 miles from Arica, might have imagined themselves safe from any effects which could be produced by an earthquake taking place so far away from them. But on the night between August 13th and 14th, the sea around this island-group rose in a surprising manner, insomuch that many thought the islands were sinking, and would shortly subside altogether beneath the waves. Some of the smaller islands, indeed, were for a time completely submerged. Before long, however, the sea fell again, and as it did so the observers "found it impossible to resist the impression that the islands were rising bodily out of the water." For no less than three days this strange oscillation of the sea continued to be experienced, the most remarkable ebbs and floods being noticed at Honolulu, on the island of Woahoo.

But the sea-wave swept onward far beyond these islands.

At Yokohama, in Japan, more than 10,500 miles from Arica, an enormous wave poured in on August 14th, but at what hour we have no satisfactory record. So far as distance is concerned, this wave affords most surprising evidence of the stupendous nature of the disturbance to which the waters of the Pacific Ocean had been subjected. The whole circumference of the earth is but 25,000 miles, so that this wave had travelled over a distance considerably greater than two-fifths of the earth's circumference. A distance which the swiftest of our ships could not traverse in less than six or seven weeks had been swept over by this enormous undulation in the course of a few hours.

More complete details reach us from the Southern Pacific.

Shortly before midnight the Marquesas Isles and the low-lying Tuamotu group were visited by the great wave, and some of these islands were completely submerged by it. The lonely Opara Isle, where the steamers which run between Panama and New Zealand have their coaling-station, was visited at about half-past eleven in the evening by a billow which swept away a portion of the coal-depot. Afterward great waves came rolling in at intervals of about twenty minutes, and several days elapsed before the sea resumed its ordinary ebb and flow.

It was not until about half-past two on the morning of August 14th, that the Samoa Isles (sometimes called the Navigator Islands) were visited by the great wave. The watchmen startled the inhabitants from their sleep by the cry that the sea was about to overwhelm them; and already, when the terrified people rushed from their houses, the sea was found to have risen far above the highest water-mark. But it presently began to sink again, and then commenced a series of oscillations, which lasted for several days, and were of a very remarkable nature. Once in every quarter of an hour the sea rose and fell, but it was noticed that it rose twice as rapidly as it sank. This peculiarity is well worth remarking. The eminent physicist Mallet speaks thus (we follow Lyell's quotation) about the waves which traverse an open sea: "The great sea-wave, advancing at the rate of several miles in a minute, consists, in the deep ocean, of a long, low swell of enormous volume, having an equal slope before and behind, and that so gentle that it might pass under a ship without being noticed. But when it reaches the edge of soundings, its front slope becomes short and steep, while its rear slope is long and gentle." On the shores visited by such a wave, the sea would appear to rise more rapidly than it sank. We have seen that this happened on the shores of the Samoa group, and therefore the way in which the sea rose and fell on the days following the great earthquake gave significant evidence of the nature of the sea-bottom in the neighborhood of these islands. As the change of the great wave's figure could not have been quickly communicated, we may conclude with certainty that the Samoan Islands are the summits of lofty mountains, whose sloping sides extend far toward the east.

This conclusion affords interesting evidence of the necessity of observing even the seemingly trifling details of important phenomena.

The wave which visited the New-Zealand Isles was altogether different in character, affording a noteworthy illustration of another remark of Mallet's. He says that where the sea-bottom slopes in such a way that there is water of some depth close in-shore, the great wave may roll in and do little damage; and we have seen that so it happened in the case of the

Samoan Islands. But he adds, that "where the shore is shelving, there will be first a retreat of the water, and then the wave will break upon the beach and roll far in upon the land." This is precisely what happened when the great wave reached the eastern shores of New Zealand, which are known to shelve down to very shallow water continuing far away to sea toward the east.

At about half-past three on the morning of August 14th, the water began to retreat in a singular manner from the port of Littleton, on the eastern shores of the southernmost of the New-Zealand Islands. At length the whole port was left entirely dry, and so remained for about twenty minutes. Then the water was seen returning like a wall of foam ten or twelve feet in height, which rushed with a tremendous noise upon the port and town. Toward five o'clock the water again retired, very slowly as before, not reaching its lowest ebb until six. An hour later, a second huge wave inundated the port. Four times the sea retired and returned with great power at intervals of about two hours. Afterward the oscillation of the water was less considerable, but it had not wholly ceased until August 17th, and only on the 18th did the regular ebb and flow of the tide recommence.

Around the Samoa group the water rose and fell once in every fifteen minutes, while on the shores of New Zealand each oscillation lasted no less than two hours. Doubtless the different depths of water, the irregular conformation of the island-groups, and other like circumstances, were principally concerned in producing these singular variations. Yet they do not seem fully sufficient to account for so wide a range of difference. Possibly a cause yet unnoticed may have had something to do with the peculiarity. In waves of such enormous extent, it would be quite impossible to determine whether the course of the wave-motion was directed full upon a line of shore or more or less obliquely. It is clear that in the former case the waves would seem to follow each other more swiftly than in the latter, even though there were no difference in their velocity.

Far on beyond the shores of New Zealand the great wave coursed, reaching at length the coast of Australia. At dawn of August 14th, Moreton Bay was visited by five well-marked waves. At Newcastle, on the Hunter River, the sea rose and fell several times in a remarkable manner, the oscillatory motion commencing at half-past six in the morning. But the most significant evidence of the extent to which the sea-wave travelled in this direction was afforded at Port Fairy, Belfast, South Victoria. Here the oscillation of the water was distinctly perceived at mid-day on August 14th; and yet, to reach this point, the sea-wave must not only have travelled on a circuitous course nearly equal in length to half the circumference of the earth, but must have passed through Bass's Straits, between Australia and Van Diemen's Land, and so have lost a considerable portion of its force and dimensions. When we remember that had not the effects of the earth-shock on the water been limited by the shores of South America, a wave of disturbance equal in extent to that which travelled westward would have swept toward the east, we see that the force of the shock was sufficient to have disturbed the waters of an ocean covering the whole surface of the earth. For the sea-waves which reached Yokohama in one direction and Port Fairy in another had each traversed a distance nearly equal to half the earth's circumference; so that if the surface of the earth were all sea, waves setting out in opposite directions from the centre of disturbance would have met each other at the antipodes of their starting-point.

It is impossible to contemplate the effects which followed the great earthquake—the passage of a sea-wave of enormous volume over fully one-third of the earth's surface, and the force with which, on the farthermost limits of its range, the wave rolled in upon shores more than 10,000 miles from its starting-place—without feeling that those geologists are right who deny that the subterranean forces of the earth are diminishing in intensity. It may be difficult, perhaps, to look on the effects which are ascribed to ancient earth-throes without imagining for a while that the power of modern earthquakes is altogether less. But when we consider fairly the share which time had in those ancient processes of change, when we see that while mountain-ranges were being upheaved or valleys depressed to their present position, race after race and type after type appeared on the earth, and lived out the long lives which belong to races and to types, we are recalled to the remembrance of the great work which the earth's subterranean forces are still engaged upon. Even now, continents are being slowly depressed or upheaved; even now mountain-ranges are being raised to a new level, table-lands are in process of formation, and great valleys are being gradually scooped out. It may need an occasional outburst such as the earthquake of August, 1868, to remind us that great forces are at work beneath the earth's surface. But, in reality, the signs of change have long been noted. Old shore-lines shift their place, old soundings vary; the sea advances in one place and retires in another; on every side Nature's plastic hand is at work modelling and remodelling the earth, in order that it may always be a fit abode for those who are to dwell upon it.

(From *Fraser's Magazine*, July, 1870.)

---

## *THE USEFULNESS OF EARTHQUAKES.*

We have lately had fearful evidence of the energy of the earth's internal forces. A vibration which, when considered with reference to the dimensions of the earth's globe, may be spoken of as an indefinitely minute quivering limited to an insignificant area, has sufficed to destroy the cities and villages of whole provinces, to cause the death of thousands of human beings, and to effect a destruction of property which must be estimated by millions of pounds sterling. Such a catastrophe as this serves indeed to show how poor and weak a creature man is in presence of the grand workings of Nature. The mere throes which accompany her unseen subterranean efforts suffice to crumble man's strongest buildings in a moment into dust, while the unfortunate inhabitants are either crushed to death among the ruins, or forced to remain shuddering spectators of the destruction of their homes.

At first sight it may seem paradoxical to assert that earthquakes, fearfully destructive as they have so often proved, are yet essentially preservative and restorative phenomena; yet this is strictly the case. Had no earthquakes taken place in old times, man would not now be living on the face of the earth; if no earthquakes were to take place in future, the term of man's existence would be limited within a range of time far less than that to which it seems likely, in all probability, to be extended.

If the solid substance of the earth formed a perfect sphere in ante-geologic times—that is, in ages preceding those to which our present geologic studies extend—

there can be no doubt that there was then no visible land above the surface of the water; the ocean must have formed a uniformly deep covering to the submerged surface of the solid globe. In this state of things, nothing but the earth's subterranean forces could tend to the production of continents and islands. Let us be understood. We are not referring to the possibility or impossibility that lands and seas should suddenly have assumed their present figure without convulsion of any sort; this *might* have happened, since the Creator of all things can doubtless modify all things according to His will; we merely say that, assuming that in the beginning, as now, He permitted all things to work according to the laws He has appointed, then, undoubtedly, the submerged earth must have risen above the sea by the action of those very forms of force which produce the earthquake in our own times.

However this may be, it is quite certain that when once continents and islands had been formed, there immediately began a struggle between destructive and restorative (rather, perhaps, than preservative) forces.

The great enemy of the land is water, and water works the destruction of the land in two principal ways.

In the first place, the sea tends to destroy the land by beating on its shores, and thus continually washing it away. It may seem at first sight that this process must necessarily be a slow one; in fact, many may be disposed to say that it is certainly a slow process, since we see that it does not alter the forms of continents and islands perceptibly in long intervals of time. But, as a matter of fact, we have never had an opportunity of estimating the full effects of this cause, since its action is continually being checked by the restorative forces we shall presently have to consider. Were it not thus checked, there can be little doubt that its effects would be cumulative; for the longer the process continued—that is, the more the land was beaten away—the higher would the sea rise, and the greater power would it have to effect the destruction of the remaining land.

We proceed to give a few instances of the sea's power of effecting the rapid destruction of the land when nothing happens to interfere with the local action—premising, that this effect is altogether insignificant in comparison with that which would take place, even in that particular spot, if the sea's action were *everywhere* left unchecked.

The Shetland Isles are composed of substances which seem, of all others, best fitted to resist the disintegrating forces of the sea — namely, granite, gneiss, mica-slate, serpentine, greenstone, and many other forms of rock; yet, exposed as these islands are to the uncontrolled violence of the Atlantic Ocean, they are undergoing a process of destruction which, even within historical times, has produced very noteworthy changes. "Steep cliffs are hollowed out," says Sir Charles Lyell, "into deep caves and lofty arches; and almost every promontory ends in a cluster of rocks, imitating the forms of columns, pinnacles, and obelisks." Speaking of one of the islands of this group, Dr. Hibbert says: "The isle of Stenness presents a scene of unequalled desolation. In stormy winters, large blocks of stone are overturned, or are removed from their native beds, and hurried to a distance almost incredible. In the winter of 1802, a tabular mass, eight feet two inches by seven feet, and five feet one inch thick, was dislodged from its bed, and carried to a distance of from eighty to ninety feet." In other parts of the Shetland Isles, where the sea has encountered less solid materials, the work of destruction has proceeded yet more effectively. In Roeness, for example, the sea has wrought its way so fiercely, that a large cavernous aperture 250 feet long has been hollowed out. "But the most sublime scene," says Dr. Hibbert, "is where a mural pile of porphyry, escaping the process of disintegration that is devastating the coast, appears to have been left as a sort of rampart against the inroads of the ocean. The Atlantic, when provoked by wintry gales, batters against it with all the force of real artillery; and the waves, in their repeated assaults, have at length forced for themselves an entrance. This breach, named the

Grind of the Navir, is widened every winter by the overwhelming surge that, finding a passage through it, separates large stones from its sides, and forces them to a distance of no less than 180 feet. In two or three spots, the fragments which have been detached are brought together in immense heaps, that appear as an accumulation of cubical masses, the product of some quarry."

Let us next turn to a portion of the coast-line of Great Britain which is neither defended, on the one hand, by barriers of rock, nor attacked, on the other, by the full fury of the Atlantic currents. Along the whole coast of Yorkshire, we find evidences of a continual process of dilapidation. Between the projecting headland of Flamborough and Spurn Point (the coast of Holderness), the waste is particularly rapid. Many spots, which are now mere sand-banks, are marked in the old maps of Yorkshire as the sites of ancient towns and villages. Speaking of Hyde (one of these), Pennant says: "Only the tradition is left of this town." Owthorne and its church have been for the most part destroyed, as also Auburn, Hartburn, and Kilnsea. Mr. Phillips, in his "Geology of Yorkshire," states that not unreasonable fears are entertained that, at some future time, Spurn Point itself will become an island, or be wholly washed away, and then the ocean, entering into the estuary of the Humber, will cause great devastation. Pennant states that "several places, once towns of note upon the Humber, are now only recorded in history; and Ravensperg was at one time a rival of Hull, and a port so very considerable in 1342, that Edward Baliol and the confederated English barons sailed from hence to invade Scotland; and Henry IV., in 1399, made choice of this port to land at, to effect the deposal of Richard II.; yet the whole of this has since been devoured by the merciless ocean; extensive sands, dry at low water, are to be seen in their stead." The same writer also describes Spurn Point as shaped like a sickle, and the land to the north, he says, was "perpetually preyed on by the fury of the German Sea, which devours whole acres at a time."

The decay of the shores of Norfolk and Suffolk is also remarkably rapid. Sir Charles Lyell relates some facts which throw an interesting light on the ravages which the sea commits upon the land here. It was computed that when a certain inn was built at Sherringham, seventy years would pass before the sea could reach the spot; "the mean loss of land being calculated from previous observations to be somewhat less than one yard annually." But no allowance had been made for the fact that the ground sloped *from* the sea. In consequence of this peculiarity, the waste became greater and greater every year as the cliff grew lower. "Between the years 1824 and 1829, no less than seventeen yards were swept away;" and when

Sir Charles Lyell saw the place, only a small garden was left between the building and the sea. We need hardly add that all vestiges of the inn have long since been swept away. Lyell also relates that, in 1829, there was a depth of water sufficient to float a frigate at a point where, less than half a century before, there stood a cliff fifty feet high with houses upon it.

We have selected these portions of the coast of Great Britian, not because the destruction of our shores is greater here than elsewhere, but as serving to illustrate processes of waste and demolition which are going on around all the shores, not merely of Great Britain, but of every country on the face of the earth. Here and there, as we have said, there are instances in which a contrary process seems to be in action. Low-lying banks and shoals are formed—sometimes along stretches of coast extending for a considerable distance. But when we consider these formations closely, we find that they rather afford evidence of the energy of the destructive forces to which the land is subject than promise to make up for the land which has been swept away. In the first place, every part of these banks consists of the *débris* of other coasts. Now, we cannot doubt that of earth which is washed away from our shores, by far the larger part finds its way to the bottom of the deep seas; a small proportion only can be brought (by some peculiarity in the distribution of ocean-currents, or in the progress of the tidal wave) to aid in the formation of shoals and banks. The larger, therefore, such shoals and banks may be, the larger must be the amount of land which has been washed away never to reappear. And although banks and shoals of this sort grow year by year larger and larger, yet (unless added to artificially) they continue always either beneath the surface of the water, in the case of shoals, or but very slightly raised above the surface. Now, if we suppose the destruction of land to proceed unchecked, it is manifest that at some period, however remote, the formation of shoals and banks must come to an end, owing to the continual diminution of the land from the demolition of which they derive their substance. In the mean time, the bed of the sea would be continually filling up, the level of the sea would be continually rising, and thus the banks would either be wholly submerged through the effect of this cause alone, or they would have so slight an elevation above the sea-level, that they would offer little resistance to the destructive effects of the sea, which would now have no other land to act upon.

But we have yet to consider the second principal cause of the wasting away of the land. The cause we have just been dealing with acts upon the shores or outlines of islands and continents; the one we have now to consider acts upon their interior. It will, perhaps, hardly be supposed that the fall of rain upon the land could have any appreciable influence in the demolition of continents; but as a matter of fact, there are few causes to which geologists are disposed to ascribe more importance. The very fact that enormous deltas have been formed at the mouths of many rivers—in other words, the actual growth of continents through the effects of rainfall—is a proof how largely this cause must tend to destroy and disintegrate the interiors of our continents. Dwelling on this point, Sir Charles Lyell presents the following remarkable illustration: "During a tour in Spain," he writes, "I was surprised to see a district of gently-undulating ground in Catalonia, consisting of red and gray sandstone, and in some parts of red marl, almost entirely denuded of herbage; while the roots of the pines, holm oaks, and some other trees, were half exposed, as if the soil had been washed away by a flood. Such is the state of the forests, for example, between Oristo and Vich, and near San Lorenzo. But, being overtaken by a violent thunderstorm in the month of August, I saw the whole surface, even the highest levels of some flat-topped hills, streaming with mud, while on every declivity the devastation of torrents was terrific. The peculiarities in the physiognomy of the district were at one explained; and I was taught that, in speculating on the greater effects which the direct action of rain may once have produced on the surface of certain parts of England, we need not revert to periods when the heat of the climate was tropical."

Combining the effects of the sea's action upon the shores of continents, and of the action of rain upon their interior, and remembering that unless the process of demolition were checked in some way, each cause would act from year to year with new force—one through the effects of the gradual rise of the sea-bed, and the other through the effects of the gradual increase of the surface of ocean exposed to the vaporizing action of the sun, which increase would necessarily increase the quantity of rain yearly precipitated on the land—we see the justice of the opinion expressed by Sir John Herschel, that, "had the primeval world been constructed as it now exists, time enough has elapsed, and force enough directed to that end has been in activity, to *have long ago destroyed every vestige of land.*"

We see, then, the necessity that exists for the action of some restorative or preservative force sufficient to counteract the effects of the continuous processes of destruction we have indicated above. If we consider, we shall see that the destructive forces owe their efficiency to their levelling action, that is, to their influence in reducing the solid part of the earth to the figure of a perfect sphere; therefore the form of force which is required to counteract them is one that shall tend to produce irregularities in the surface-contour of the earth. And it will be remarked that, although *upheaval* is the process which appears at first sight to be the only effectual remedy to the levelling action of rains and ocean-currents, yet the forcible depression of the earth's surface may prove in many instances yet more effective, since it may serve to reduce the sea-level in other places.

Now, the earth's subterranean forces serve to produce the very effects which are required in order to counteract the continual disintegration of the shores and interior parts of continents. In the first place, their action is not distributed with any approach to uniformity over different parts of the earth's crust, and therefore the figure they tend to give to the surface of that crust is not that of a perfect sphere. This, of itself, secures the uprising of some parts of the solid earth above the sea-level. But this is not all. On a comparison of the various effects due to the action of subterranean forces, it has been found that the forces of upheaval act (on the whole) more powerfully under continents, and especially under the shore-lines of continents, while the forces of depression act most powerfully (on the whole) under the bed of the ocean. It need hardly be said that whenever the earth is upheaved in one part, it must be depressed somewhere else. Not necessarily at the same instant, it should be remarked. The process of upheaval may be either momentarily accompanied by a corresponding process of depression, or the latter process may take place by a gradual action of the elastic powers of the earth's crust; but in one way or the other, the balance between upheaval and depression must be restored. Hence, if it can be shown that for the most part the forces of upheaval act underneath the land, it follows—though we may not be able to recognize the fact by obvious visible signs—that processes of depression are taking place underneath the ocean. Now, active volcanoes mark the centre of a district of upheaval, and nearly all volcanoes are found near the sea. It seems as if Nature had provided against the inroads of the ocean by seating the earth's upheaving forces just where they are most wanted.

Even in earthquake districts which have no active vent, the same law is found to prevail. It is supposed by the most eminent seismologists that earthquake regions around a volcano, and earthquake regions apparently disconnected form any outlet, differ only in this respect, that in the one case the subterranean forces have had sufficient power to produce the phenomena of eruption, while in the other they have not. In "earthquakes," says Humboldt, "we have evidence of a volcano-producing force; but such a force, as universally diffused as the internal heat of the globe, and proclaiming itself everywhere, rarely acts with sufficient energy to produce actual eruptive phenomena; and when it does so, it is only in isolated and particular places."

Of the influence of the earth's subterranean forces in altering the level of the land, we might quote many remarkable instances, but considerations of space compel us to confine ourselves to two or three. The slow processes of upheaval or depression may, perhaps, seem less immediately referable to subterranean action than those which are produced during the progress of an actual earthquake. We pass over, therefore, such phenomena as the gradual uprising of Sweden, the slow sinking of Greenland, and (still proceeding westward) the gradual uprising of Nova Scotia and the shores of Hudson's Bay. Remarkable and suggestive as these phenomena really are, and indisputable as the evidence is on which they rest, they will probably seem much less striking to our readers than those which we are now about to quote.

On the 19th of November, 1822, a widely-felt and destructive earthquake was experienced in Chili. On the next day, it was noticed for the first time that a broad line of sea-coast had been deserted by the sea for more than one hundred miles. A large part of this tract was covered by shell-fish, which soon died, and exhaled the most offensive effluvia. Between the old low-water mark and the new one, the fishermen found burrowing shells, which they formerly had to search for amid the surf. Rocks some way out to sea which had formerly been covered, were now dry at half ebb-tide.

Careful measurements showed that the rise of the land was greater at some distance inshore than along the beach. The water-course of a mill about a mile inland from the sea had gained a fall of fourteen inches in little more than a hundred yards. At Valparaiso, the rise was three feet; at Quintero, four feet.

In February, 1835, and in November, 1837, a large tract of Chili was similarly shaken, a permanent rise of two feet following the former earthquake, and a rise of eight feet the latter.

The earthquake which took place at Cutch in 1819 is perhaps in some respects yet more remarkable. In this instance, phenomena of subsidence, as well as phenomena of upheaval, were witnessed. The estuary of the Indus, which had long been closed to navigation—being, in fact, only a foot deep at ebb-tide, and never more than six feet at flood—was deepened in parts to more than eighteen feet at low water. The fort and village of Sindree was submerged, only the tops of houses and walls being visible above the water. But although this earthquake seemed thus to have a land-destroying instead of a land-creating effect, yet the instances of upheaval were, even in this case, far more remarkable than those of depression. "Immediately after the shock," says Sir Charles Lyell, "the inhabitants of Sindree saw at a distance of five miles and a half from their village a long, elevated mound, where previously there had been a low and perfectly level plain. To this uplifted tract they gave the name of Ullah-Bund, or the 'Mound of God,' to distinguish it from several artificial dams previously thrown across the eastern arm of the Indus. It has been ascertained," he adds, "that this new-raised country is upward of fifty miles in length from east to west, running parallel to the line of subsidence which caused the grounds around Sindree to be flooded. The breadth of the elevation is conjectured to be in some parts sixteen miles, and its greatest ascertained height above the original level of the delta is ten feet—an elevation which appears to the eye to be very uniform throughout."

(From *Chambers's Journal*, November 7, 1868.)

---

## *THE FORCING POWER OF RAIN.*

There is an old proverb which implies that England need never fear drought; and we have had clear evidence this year that an exceptionally dry summer is not necessarily followed by a bad harvest. But we believe that when a balance is carefully struck between the good and the evil effects resulting from excessive drought in England, it will be found that the latter largely prevail. In fact, it is only necessary to observe the effects which have followed the recent wet weather to recognize the fact that rain has a forcing power, the very diminished supply of which at the due season cannot fail to have seriously injurious effects. In various parts of England we see evidences of the action of such a power during the present autumn in the blossoming of trees, in the flowering of primroses and other spring plants, in rich growths of fungi, and in various other ways. It cannot be doubted that there is here a comparative waste of powers, which, expended in due season, would have produced valuable results.

The modern theories of the correlation of force suffice to show how enormous a loss a country suffers when there is a failure in the supply of rain, or when that supply comes out of its due season. When we consider rain in connection with the causes to which it is due, we begin to recognize the enormous amount of power of which the ordinary rainfall of a country is the representative; and we can well understand how it is that "the clouds drop fatness on the earth."

The sun's heat is, of course, the main agent—we may almost say the only agent—in supplying the rain-fall of a country. The process of evaporation carried on over large portions of the ocean's surface is continually storing up enormous masses of water in the form of invisible aqueous vapor, ready to be transformed into cloud, then wafted for hundreds of miles across seas and continents, to be finally precipitated over this or that country, according to the conditions which determine the downfall of rain. These processes do not appear, at first sight, indicative of any very great expenditure of force, yet, in reality, the force-equivalent of the rain-supply of England alone for a single year is something positively startling. It has been calculated that the amount of heat required to evaporate a quantity of water which would cover an area of 100 miles to a depth of one inch would bê equal to the heat which would be produced by the combustion of half a million tons of coals. The amount of force of which this consumption of heat would be the equivalent, corresponds to that which would be required to raise a weight of upward of one thousand millions of tons to a height of one mile. Now, when we remember that the area of Great Britain and Ireland is about 120,000 square miles, and that the annual rainfall averages about 25 inches, we see that the force-equivalent of the rainfall is enormous. All the coal which could be raised from our English coal-mines in thousands of years would not give out heat enough to produce England's rain-supply for a single year. When to this consideration we add the circumstance that the force of rain produces bad as well as good effects—the former when the rain falls at undue seasons or in an irregular manner, the latter only when the rainfall is distributed in the usual manner among the seasons—we see that an important loss accrues to a country in such exceptional years as the present.

There are few subjects more interesting than those depending on the correlation of physical forces; and we may add that there are few the study of which bears more largely on questions of agricultural and commercial economy. It is only of late years that the silent forces of Nature—forces continually in action, but which are too apt to pass unnoticed and unrecognized—have taken their due place in scientific inquiry. Strangely enough, the subject has been found to have at once a most practical bearing on business relations, and an aspect more strikingly poetical than any other subject, perhaps, which men of science have ever taken in hand to investigate. We see the ordinary processes of Nature, as they are termed, taking their place in the workshop of modern wealth, and at the same time exhibited in a hundred striking and interesting physical relations. What, for instance, can be stranger or more poetical than the contrast which Professor Tyndall has instituted between that old friend to the agriculturist—the wintry snow-flake—and the wild scenery of the Alps? "I have seen," he says, "the wild stone-avalanches of the Alps, which smoke and thunder down the declivities with a vehemence almost sufficient to stun the observer. I have also seen snow-flakes descending so softly as not to hurt the fragile spangles of which they were composed; yet to produce from aqueous vapor a quantity which a child could carry of that tender material demands an exertion of energy competent to gather up the shattered blocks of the largest stone-avalanche I have ever seen, and pitch them to twice the height from which they fell."

We may point out in this place the important connection which exists between the rainfall of a country and the amount of forest-land. We notice that in parts of America attention is being paid—with markedly good results—to the influence of forests in encouraging rainfall. We have here an instance in which cause and effect are interchangeable. Rain encourages the growth of an abundant vegetation, and abundant vegetation in turn aids to produce a state of the superincumbent atmosphere which encourages the precipitation of rain. The consequence is, that it is very necessary to check, before it is too late, the processes which lead to the gradual destruction of forests. If these processes are continued until the climate has become excessively dry, it is almost impossible to remedy the mischief, simply because the want of moisture is destructive to the trees which may be planted to encourage rainfalls. Thus, there are few processes more difficult (as has been found by experience in parts of Spain and elsewhere) than the change of an arid region into a vegetation-covered district. In fact, if the region is one of great extent, the attempt to effect such a change is a perfectly hopeless one. On the other hand, the contrary process—that is, the attempt to change a climate which is too moist into one of less humidity—is in general not attended with much difficulty. A judicious system of clearing nearly always leads to the desired result.

The dryness of the past year has not been due to the want of moisture in the air, nor to the exceptionally unclouded condition of our skies. We believe that, on the whole, the skies have been rather more cloudy than usual this year. The fact that so little dew has fallen is a sufficient proof that the nights have been on the whole more cloudy than usual, since, as is well known, the presence of clouds, by checking the radiation of the earth's heat, prevents (or at least diminishes) the formation of dew. The fact would seem to be that the westerly and southwesterly winds which usually blow over England during a considerable part of the year, bringing with them large quantities of aqueous vapor from above the great Gulf Stream, have this year blown somewhat higher than usual. Why this should be it is not very easy to say. The height of the vapor-laden winds is usually supposed to depend on the heat of the weather. In summer, for instance, the clouds range higher, and therefore travel farther inland before they fall in rain. In winter, on the contrary, they travel lower, and hence the rain falls more freely in the western than in the eastern countries during winter. A similar relation prevails in the Scandinavian peninsula—Norway receiving more rain in winter than in summer, while Sweden receives more rain in summer than in the winter. But this summer the rain-clouds have blown so much higher than usual as to pass beyond England altogether. Possibly we may find an explanation in the fact that before reaching our shores at all the clouds were relieved by heavy rainfalls—probably due to some exceptional electrical relations—over parts of the Atlantic Ocean. It is stated that the steamships from America this summer were, in many instances, drenched by heavy showers until they neared the coasts of England.

(From the *Daily News*, October 5, 1868.)

---

## *A SHOWER OF SNOW-CRYSTALS.*

YESTERDAY morning a remarkably fine fall of snow-stars took place over many parts of London. The crystals were larger and more perfectly formed than is commonly the case in our latitudes, where the conditions requisite for the formation of these beautiful objects are less perfectly fulfilled than in more northerly regions. Many forms were to be noticed which the researches of Scoresby, Glaisher, and Lowe, have shown to be somewhat uncommon.

Many of our readers will, perhaps, be surprised to learn that no less than 1,000 different kinds of snow-crystals have been noticed by the observers named above, and that a large proportion of them have been figured and described. The patterns are of wonderful beauty. A strange circumstance connected with these objects is the fact that for the most part they are found, on a close examination, to be formed of minute colored crystals—some red, some green, others blue or purple. In fact, all the colors of the rainbow are to be seen in the delicate tracery of these fine hexagonal stars. So that in the perfect whiteness of the driven snow we have an illustration of the well-known fact that the colors of the rainbow combine to form the purest white. For the common snow-flake is formed of a large number of such tiny crystals as were falling yesterday; though their beauty is destroyed in the snow-flake, through the effects of collision and partial melting. It may not be very commonly known that ordinary ice, also, is composed of a combination of crystals presenting all the regularity of formation seen in the snow-crystals. This would scarcely be believed by any one who examined a rough mass of ice taken from the surface of a frozen lake. Yet, if a slice be cut from the mass and placed in the sun's light, or before a fire, the beautiful phenomena called ice-flowers make their appearance—"A fairy seems to have breathed upon the ice, and caused transparent flowers of exquisite beauty suddenly to blossom in myriads within it."

When we remember that the enormous ice-bergs of the Arctic and Antarctic seas, the snow-caps which crown the Alps, and Andes, and Himalayas, and the glaciers which urge their way with resistless force down the mountain valleys, are all made up of these delicate and beautiful snow-flowers, we are struck with the force of the strange contrasts which Nature presents to our contemplation. We may say of the snow-crystals what Tennyson said of the small sea-shell. Each snow-star is

> "Frail, but a work divine,
> Made so fairily well,
> So exquisitely minute,
> A miracle of design."

Yet—massed together with all the prodigality of Nature's unsparing hand—they crown the everlasting hills; or, falling in avalanche and glacier, overwhelm the stoutest works of man; or, in vast islands of floating ice, show themselves to be

> "Of force to withstand, year upon year, the shock
> Of cataract seas that snap the three-decker's oaken spine."

(From the *Daily News*, March 11, 1869.)

---

## *LONG SHOTS.*

Our artillerists have paid more attention of late years to the destructive properties of various forms of cannon than to the question of range. It was different when first the rifling of cannon was under discussion. Then the subject which was most attentively considered (after accuracy of fire) was the range which might possibly be obtained by various improvements in the structure of rifled cannon. Many of our readers will remember how, soon after the construction of

Armstrong guns had been commenced in the Government factories, a story was spread abroad of the wonderful practice which had been made with this gun at a range of seven miles. At that tremendous range, a shot had been fired into the middle of a flock of geese, according to one version of the story; but this was presently improved upon, and we were told that a bird had been singled out of the flock by the artillerists and successfully "potted!" Many believed this little narrative; though some few, influenced perhaps by the consideration that a flock of geese would not be visible at a distance of seven miles, were obstinately incredulous. Presently it turned out that the Armstrong gun was incapable of throwing a shot to a distance of seven miles; so that a certain air of improbability has since attached to the narrative. Still there were not wanting those who referred to "Queen Anne's pocket-pistol"—the cannon which was able to throw shot across the Straits of Dover; and in the fulness of their faith in that mythical piece of ordnance, they refused to believe that the skill of modern artillerists was unequal to the construction of cannon even more effective.

If there are any who still believe in the powers ascribed to the far-famed "pocket-pistol," they will find their confidence in modern artillery largely shaken by the announcement that it is considered a great matter that one of Whitworth's cannon should have thrown a shot to a distance of very nearly six and a half miles. Not only is this so, however, but it is well known that no piece of ordnance has ever flung a projectile to so great a distance since first fire-arms were invented; and it may be safely predicted that men will never be able to construct a cannon which—as far as range is concerned—will do much better than this one of Mr. Whitworth's. The greatest range which had ever before been obtained fell somewhat short of six miles. The 7-inch steel gun contrived by Mr. Lynall Thomas had flung a projectile weighing one hundred and seventy-five pounds to a distance of ten thousand and seventy-five yards; and, according to General Lefroy's "Handbook of Artillery," that was the greatest range ever recorded. But Mr. Whitworth's cannon throws a shot more than a thousand yards farther.

Very few have any idea of the difficulties which oppose themselves to the attainment of a great range in artillery practice. It may seem, at first sight, the simplest possible matter to obtain an increase of range. Let the gun be made but strong enough to bear a sufficient charge, and range seems to be merely a question of the quantity of powder made use of. But in reality the matter is much more complicated. The artillerist has to contrive that the whole of the powder made use of shall be burned before the shot leaves the cannon, and yet that the charge shall not explode so rapidly as to burst the cannon. If he used some forms of powder, very useful for special purposes, half the charge would be blown out without doing its share of work. On the other hand, there are some combustibles—as gun-cotton and the nitrates—which burn so fast that the gun would be likely to burst before the shot could be expelled. Then, again, the shot must fit so closely that there shall be no windage, and yet not so closely as to resist too much the action of the exploding powder. Again, there is the form of the shot to be considered. A sphere is not the solid which passes most readily through a resisting medium like the air; and yet, other projectiles, which are best so long as they maintain a certain position, meet with a greater resistance when once they begin to move unsteadily. The conoid used in ordinary rifle-practice, for example, passes much more freely through the air, point first, than an ordinary spherical bullet; but if the point did not travel first, as would happen but for the rifling, or even if the conoidal bullet "swayed about" on its course, it would meet with more resistance than a spherical bullet. Hence the question of "fast or slow rifling" has to be considered. "Fast rifling" gives the greater spin, but causes more resistance in the exit of the shot from the barrel; with "slow rifling," these conditions are reversed.

And then the common notion is that a cannon-ball travels in the curve called a parabola, and that artillerists have nothing to do but to calculate all about this parabola, and to deduce the range from the initial velocity according to some simple principles deduced from the properties of the curve. All this is founded on a complete misapprehension of the true difficulties in the way of the problem. Only projectiles thrown with small velocity from the earth travel in parabolic paths. A cannon-ball follows a wholly different kind of curve. The resistance of the air, which seems to most persons a wholly insignificant item in the inquiry, is so enormous in the case of a cannon-ball as to become by far the most important difficulty in the way of the practical artillerist. When a 250-pound shot is hurled with such force from a gun as to cover a range of six miles, the resistance of the air is about forty times the weight of the ball—that is, is equivalent to a weight of upward of four tons. The range is such a case as this is but a small fraction of that which would be given by the ordinary parabolic theory.

As regards artillery practice in war, there are other difficulties in the attainment of a very extended range. Cannon meant for battering down forts cannot possibly be used in the same way that Whitworth's was used at Shoeburyness. If the shot flung from this gun at an elevation of thirty-three degrees could have been watched, it would have been found that it fell to the earth at a much greater angle—that is, much more nearly in a perpendicular direction. On the ordinary parabolic theory, of course, the angle of fall would be the same as the angle of elevation, but under actual circumstances there is an important difference. If forts are to be battered down, however, it will not serve that they should be struck from above; our artillerists must perforce keep to the old method of pounding away at the face of the forts they attack. Therefore, an elevation which is all very well for mortars—that is, when the question merely is of flinging a bomb into a town or fortress—is utterly unsuited for ordinary artillery. With an elevation of 10°, Whitworth's cannon scarcely projected the 250-lb. shot to a distance of three miles.

The progress of the modern science of gunnery certainly tends to increase the distance at which armies will engage each other. With field artillery flinging shot to a distance of two or three miles, and riflemen able to make tolerably sure practice at a distance of three-quarters of a mile, we are not likely often to hear of hand-to-hand conflicts in future warfare. The use of breech-loaders will also tend to the same effect. Hitherto we have scarcely had experience of the results which these changes are to produce on modern warfare. At Sadowa breech-loaders did not encounter breech-loaders, and it was easy for the victors in that battle to come to close quarters with their enemies. But in a battle where both sides are armed with breech-loaders, we shall probably see another sort of affair altogether. The bayonet will be an almost useless addition to the soldier's arms; a charge of cavalry upon well-armed infantry will be almost as hopeless as the famous Balaklava charge; and the artillery on either side will have to play a game at long shots. We venture to anticipate that the first great European war will introduce a total change into the whole system of warlike manœuvres.*

(From the *Daily News*, November, 1868.)

---

## *INFLUENCE OF MARRIAGE ON THE DEATH-RATE.*

The Royal Commission on the Law of Marriage has attracted attention to many singular and instructive results of modern statistical inquiry. Not the least important of these is the apparent influence of marriage on the death-rate. For several years it has been noticed by statisticians that the death-rate of unmarried men is considerably higher than the death-rate of married men and widowers. We believe that Dr. Stark, Registrar-General for Scotland, was one of the first to call attention to this peculiarity, as evidenced by the results of two years' returns for Scotland. But the law has since been confirmed by a far wider range of statistical inquiry. The relative proportion between

* The reader need hardly be reminded of the most complete fulfilment of this anticipation.

the death-rates of the married and of the unmarried is not absolutely uniform in different countries, but it is fairly enough represented by the following table, which exhibits the mortality per thousand of married and unmarried men in Scotland:

| Ages. | Husbands and Widowers. | Unmarried. |
|---|---|---|
| 20 to 25 | 6.26 | 12.31 |
| 25 to 30 | 8.23 | 14.94 |
| 30 to 35 | 8.65 | 15.94 |
| 35 to 40 | 11.67 | 16.02 |
| 40 to 45 | 14.07 | 18.35 |
| 45 to 50 | 17.04 | 21.18 |
| 50 to 55 | 19.54 | 26.34 |
| 55 to 60 | 26.14 | 28.54 |
| 60 to 65 | 35.63 | 44.54 |
| 65 to 70 | 52.93 | 60.21 |
| 70 to 75 | 81.56 | 102.71 |
| 75 to 80 | 117.85 | 143.94 |
| 80 to 85 | 173.88 | 195.40 |

From this table we are to understand that out of one hundred thousand married persons (including widowers) from 20 to 25 years old, 626 die in the course of each year; while out of a similar number of unmarried persons, between the same ages, no less than 1,231 die in each year. And in like manner all the other lines of the table are to be interpreted.

Commenting on the evidence supplied by the above figures, Dr. Stark stated that "bachelorhood is more destructive to life than the most unwholesome trades, or than residence in an unwholesome house or district, where there has never been the most distant attempt at sanitary improvement of any kind." And this view has been very generally accepted, not only by the public, but by professed statisticians. Yet as a matter of fact, we believe that no such inferences can legitimately be drawn from the above table. Dr. Stark appears to us to have fallen into the mistake, which M. Quetelet tells us is so common, of trying to make his statistics carry more weight than they are capable of bearing. It is important that the matter should be put in a just light, for the Royal Commission on the Law of Marriage has revealed no more striking fact than that of the prevalence of immature marriages, and such reasoning as Dr. Stark's certainly cannot tend to discourage these unwise alliances. If death strikes down in five years only half as many of those who are married as of those who are unmarried between the age of 20 and 25 (as appears from the above table), and if the proportion of deaths between the two classes goes on continually diminishing in each successive lustre (as is also shown by the above table), it seems reasonable to infer that the death-rate would be even more strikingly disproportionate in the case of persons between the ages of fifteen and twenty than in the case of persons between the ages of twenty and twenty-five. We believe, indeed, that if Dr. Stark had extended his table to include the former ages, the result would have been such as we have indicated.

Yet few will suppose that such very youthful marriages can exercise so singularly beneficial an effect.

To many, Dr. Stark's conclusion may appear to be a natural and obvious *sequitur* from the evidence upon which it is founded. Admitting the facts—and we see no reason for doubting them—it may appear at first sight that we are bound to accept the conclusion that matrimony is favorable to longevity. Yet the consideration of a few parallel cases will suffice to show how small a foundation the figures we have quoted supply for such a conclusion. What would be thought, for example, of any of the following inferences? Among hot-house plants there are observed a greater variety and brilliancy of color than among those which are kept in the open air, therefore the housing of plants conduces to the splendor of their coloring. Or again: The average height of Life Guardsmen is greater than that of the rest of the male population, therefore to be a Life Guardsman conduces to tallness of stature. Or, to take an example still more closely illustrative of Dr. Stark's reasoning—the average longevity of noblemen exceeds that of untitled persons, therefore to have a title is conducive to longevity; or to borrow his words, "to remain without a title is more destructive to life than the most unwholesome trades, or than residence in an unwholesome house or district, where there has never been the most distant attempt at sanitary improvement of any kind."

We know that the inference is absurd in each of the above instances, and we are able at once to show where the flaw in the reasoning lies. We know that splendid flowers are more commonly selected for housing, and that Life Guardsmen are chosen for their tallness, so that we are prevented from falling into the mistake of ascribing splendor of color in the one instance, or tallness in the other, to the influence of causes which have nothing whatever to do with those attributes; nor is any one likely to ascribe the longevity of our nobility to the possession of a title. Yet there is nothing in any one of the above inferences which is in reality more unsound than Dr. Stark's inference from the mortality bills, when the latter are considered with due reference to the principles of interpretation which statisticians are bound to follow.

The fact is, that in dealing with statistics the utmost care is required in order that our inferences may not be pushed beyond the evidence afforded by our facts. In the present instance, we have simply to deal with the fact that the death-rate of unmarried men is higher than the death-rate of married men and widowers. From this fact we cannot reason as Dr. Stark has done to a *simple* conclusion. All that we can do is to show that one of *three* conclusions must be adopted: Either matrimony is favorable (directly or indirectly) to longevity, in a degree sufficient wholly to account for the observed peculiarity; or a principle of selection—the effect of which is such as, on the whole, to fill the ranks of married men from among the healthier and stronger portion of the community—operates in a sufficient degree to account wholly for the observed death-rates; or, lastly, the observed death-rates are due to the combination, in some unknown proportion, of the two causes just mentioned.

No reasonable doubt can exist, as it seems to us, that the third is the true conclusion to be drawn from the evidence supplied by the mortality bills. Unfortunately, the conclusion thus deduced is almost valueless, because we are left wholly in doubt as to the proportion which subsists between the effects to be ascribed to the two causes thus shown to be in operation. It scarcely required the evidence of statistics to prove that each cause must operate to some extent. It is perfectly obvious, on the one hand, that although hundreds of men who would be held by insurance companies to be "bad lives" may contract marriage, yet on the whole a principle of selection is in operation which must tend to bring the healthier portion of the male community into the ranks of the married, and to leave the unhealthier in the state of bachelorhood. A little consideration will show also that, on the whole, the members of the less healthy trades, very poor persons, habitual drunkards, and others whose prospects of long life are unfavorable, must (on the average of a large number) be more likely to remain unmarried than those more favorably situated. Another fact drawn from the Registrar-General's returns suffices to prove the influence of poverty on the marriage-rate. We refer to the fact that marriages are invariably more numerous in seasons of prosperity than at other times. Improvident marriages are undoubtedly numerous, but prosperity and adversity *have* their influence, and that influence not unimportant, on the marriage returns. On the other hand, it is perfectly obvious that the life of a married man is likely to be more favorable to longevity than that of a bachelor. The mere fact that a man has a wife and family depending upon him will suffice to render him more careful of his health, less ready to undertake dangerous employments, and so on; and there are other reasons which will occur to every one for considering the life of a married men better (in the sense of the insurance companies) than that of a bachelor. In fact, while we are compelled to reject Dr. Stark's statement that "bachelorhood is more destructive to life than the most unwholesome trades, or than residence in an unwholesome house or district, where there has never been the most distant attempt at sanitary improvement of any kind," we may safely accept his opinion that statistics "prove the truth of one of the first natural laws revealed to man—'It is not good that man should live alone.'" Whether the law required any proof is a question into which we need not enter.

(From the *Daily News*, October 17, 1868.)

## *THE TOPOGRAPHICAL SURVEY OF INDIA.*

At the close of the war with Tippoo Sahib, Major Lambton planned the triangulation of the country lying between Madras and the Malabar coast, a district which had been roughly surveyed during the progress of the war by Colonel Mackenzie. The Duke of Wellington gave his approval to the project, and his brother, the Governor-General of India, and Lord Clive (son of the great Clive), Governor of Madras, used their influence to aid Major Lambton in carrying out his design. The only astronomical instrument made use of by the first survey-party was one of Ramsden's zenith-sectors, which Lord Macartney had placed in the hands of Dinwiddie, the astronomer, for sale. A steel chain, which had been sent with Lord Macartney's embassy to the Emperor of China and refused, was the only apparatus available for measuring.

Thus began the Great Trigonometrical Survey of India, a work whose importance it is hardly possible to over-estimate. Conducted successively by Colonel Lambton, Sir George Everest, Sir Andrew Waugh, and Lieutenant-Colonel Walker (the present superintendent), the trigonometrical survey has been prosecuted with a skill and accuracy which render it fairly comparable with the best works of European surveyors. But to complete in this style the survey of the whole of India would be the work of several centuries. The trigonometrical survey of Great Britain and Ireland has been already more than a century in progress, and is still unfinished. It can, therefore, be imagined that the survey of India—nearly ten times the size of the British Isles, and presenting difficulties a hundred-fold greater than those which the surveyor in England has to encounter—is not a work which can be quickly completed.

But the growing demands of the public service have rendered it imperatively necessary that India should be rapidly and completely surveyed. This necessity led to the commencement of the Topographical Survey of India, a work which has been pushed forward at a surprising rate during the past few years. Our readers may form some idea of the energy with which the survey is in progress from the fact that Colonel Thuillier's Report for the season 1866–'67 announces the charting of an area half as large as Scotland, and the preparatory triangulation of an additional area nearly half as large as England.

In a period of thirty years, with but few surveying parties at first, and a slow increase in their number, an area of 160,000 square miles has been completed and mapped by the topographical department. The revenue surveyors have also supplied good maps (on a similar scale) of 364,000 square miles of country during the twenty years ending in 1866. Combining these results, we have an area of 524,000 square miles, or upward of four times that of Great Britain and Ireland. For all this enormous area the surveyors have the records in a methodical and systematic form, fit for incorporation in the atlas of India. Nor does this estimate include the older revenue surveys of the northwestern provinces, which, for want of proper supervision in former years, were never regularly reduced. The records of these surveys were destroyed in the mutiny—chiefly in Hazaumbaugh and the southwestern frontier agency. The whole of these districts remain to be gone over in a style very superior to that of the last survey.

The extent of the country which has been charted may lead to the impression that the survey is little more than a hasty reconnoissance. This, however, is very far indeed from being the case. The preliminary triangulation, which is the basis of the topographical survey, is conducted with extreme care. In the present Report, for instance, we find that the discrepancies between the common sides of the triangles—in other words, the discrepancies between the results obtained by different observers—are in some cases less than one-tenth of an inch per mile; in others they are from one inch to a foot per mile; and in the survey of the Cossyah and Garrow Hills, where observations had to be taken to large objects such as trees, rocks, etc., with no defined points for guidance, the results differ by as much as twenty-six inches per mile. These discrepancies must not only be regarded as insignificant in themselves, but must appear yet more trifling when it is remembered that they are not cumulative, inasmuch as the preliminary triangulation is itself dependent on the great trigonometrical survey.

Let us understand clearly what are the various forms of survey which are or have been in progress in India. There are three forms to be considered: (1) The Great Trigonometrical Surveys; (2) The Revenue Surveys; and (3) the Topographical Surveys.

Great trigonometrical operations are extended in a straight course from one measured base to another. Every precaution which modern skill and science can suggest is taken in the measurement of each base-line, and in the various processes by which the survey is extended from one base-line to the other. The accuracy with which work of this sort is conducted may be estimated from the following instance: During the progress of the Ordnance Survey of Great Britain and Ireland, a base-line nearly eight miles long was measured near Loch Foyle in Ireland, and another nearly seven miles long on Salisbury Plain. Trigonometrical operations were then extended from Loch Foyle to Salisbury Plain, a distance of about 340 miles; and the Salisbury base-line was calculated from the observations made over this long arc. *The difference between the measured and calculated values of the base-line was less than five inches!* As we have stated, the trigonometrical survey of India will bear comparison with the best work of our surveyors in England.

A revenue survey is prosecuted for the definition of the boundaries of estates and properties. The operations of such a survey are therefore carried on conformably to those boundaries.

The topographical survey of a country is defined by Sir A. Scott Waugh to imply "the measurement and delineation of the natural features of a country, and the works of man thereon, with the object of producing a complete and sufficiently accurate map. Being free from the trammels of boundaries of properties, the principal lines of operations must conform to the features of the country, and objects to be surveyed."

The only safe basis for the topographical survey of a country is a system of accurate triangulation. And where the extent of country to be surveyed is large, there will always be a great risk of the accumulation of error in the triangulation itself; which must therefore be made to depend on the accurate results obtained by the great trigonometrical operations. In order to secure this result, fixed stations are established in the vicinity of the great trigonometrical series. Where this plan cannot be adopted, a net-work of large symmetrical triangles is thrown over the district to be surveyed, or boundary series of triangles are carried along the outline of the district or along convenient internal lines. The former of these methods is applicable to a hilly district, the latter to a flat country.

When the district to be surveyed has been triangulated, the work of filling-in the topographical details is commenced. Each triangle being of moderate extent, with sides from three to five miles in length, and the angular points being determined, as we have seen, with great exactness, it is evident that no considerable error can occur in filling-in the details. Hence, methods can be adopted in the final topographical work which would not be suitable for triangulation. The triangles can either be "measured up," or the observer may traverse from trigonometrical point to point, taking offsets and intersections; or, lastly, he may make use of the plane-table. The first two methods require little comment; but the principle of plane-tabling enters so largely into Indian surveying, that our notice would be incomplete without a brief account of this simple and beautiful method.

The plane-table is a flat board turning on a vertical pivot. It bears the chart on which the observer is planning the country. Suppose, now, that two points A and B are determined, and that we require to mark in the position of a third point C: It is clear that if we observed with a theodolite the angles A B C and B A C, we might lay these down on the chart with a protractor, and so the position of C would be determined with an accuracy proportioned to the care with which the observations were made, and the corresponding constructions applied to the chart. But in "plane-tabling" a more direct plan is adopted. A ruler bearing sights, resembling those of a rifle, is so applied that the edge passing through the point A on the chart (the observer being situated at the real station A) passes through the point B on the chart, the line of sight passing through the real station B. The table being fixed in the position thus obtained, the ruler is next directed so that its edge passes through A while the line of sight points to C. A line is now ruled with a pencil through A toward C. In a similar manner, the table having been removed to the station B, a pencil-line is drawn through the point B on the chart toward C. The two lines thus drawn determine by their intersection the place of C on the chart.

The above is only one instance of the modes in which a plane-table can be applied; there are several others. Usually the magnetic compass is made use of to fix the position of the table in accordance with the true bearing of the cardinal points. Also the bearings of several points are taken around each station; and thus a variety of tests of the correctness of the work become applicable. Into such points as these we need not here enter. It is sufficient that our readers should have been enabled to gather the simple principles on which plane-tabling depends, and the accuracy with which (when suitable precautions are taken) it can be applied as a method of observation subsidiary to the ordinary trigonometrical processes.

"A hilly country," says Sir A. Waugh, "offers the fairest field for the practice of plane-table surveys, and the more rugged the surface the greater will be the relative advantages and facilities this system possesses over the methods of actual measurement. On the other hand, in flat lands the plane-table works at a disadvantage, while the traverse system is facilitated. Consequently, in such tracts, the relative economy of the two systems does not offer so great a contrast as in the former. In closely-wooded or jungly tracts, all kinds of survey operations are prosecuted at a disadvantage; but in such localities, the commanding points must be previously cleared for trigonometrical operations, which facilitates the use of the table."

In whatever way the topographical details have been filled-in, a rigorous system of check must be applied to the work. The system adopted is that of running lines across ground that has been surveyed.

This is done by the head of the party or by the chief assistant-surveyor. A sufficient number of points are obtained in this way for comparison with the work of the detail surveyors; and when the discrepancies exceed certain limits, the work in which they appear is rejected. Owing to the extremely unhealthy, jungly, and rugged nature of the ground in which nearly all the Indian surveys have been progressing, it has not always been found practicable to check by regularly chained lines. There are, however, other modes of testing plane-table surveys, and as these entail less labor and expense in hilly and jungly tracts, and are quite as effective if thoroughly carried out, they have been adopted generally, while the measured routes or check-lines have only been pursued under more favorable conditions. Colonel Thuillier states that "the inspection of the work of every detailed surveyor in the field has been rigorously enforced, and the work of the field season is not considered satisfactory or complete unless this duty has been attended to."

The rules laid down to insure accuracy in the survey are—first, that the greatest possible number of fixed points should be determined by regular triangulation; secondly, that the greatest possible number of plane-table fixings should be made use of within each triangle; and, lastly, that eye-sketching should be reduced to a minimum. If these rules are well attended to, the surveyor can always rely on the value of the work performed by his subordinates. But all these conditions cannot be secured in many parts of the ground allotted to the several topographical parties, owing to the quantity of forest-land and the extremely rugged nature of the country. Hence arises the necessity for test-lines to verify the details or for some rigorous system of check; and this is more especially the case where native agency is employed.

So soon as the country has been accurately planned, the configuration of the ground has to be sketched up. This process is the end and aim of all the preceding work.

The first point attended to is the arterial system, or water-drainage, constituting the outfall of the country; whence are deduced the lines of greatest depression of the ground. Next the water-sheds or ridges of hills are traced in, giving the highest level. Lastly, the minor or subordinate features are drawn in with the utmost precision attainable. "The outlines of table-land should be well defined," says Sir A. Waugh, "and ranges of hills portrayed with fidelity, carefully representing the water-sheds or *divortia aquarum*, the spurs, peaks, depressions or saddles, isthmuses or connecting-links of separate ranges, and other ramifications. The depressed points and isthmuses are particularly valuable, as being either the sites of ordinary passes or points which new roads should conform to."

And here we must draw a distinction between survey and reconnoissance. It is absolutely necessary in making a survey that the outlines of ground as defined by ridges, water-courses, and feet of hills, should be rigorously fixed by actual observation and careful measurement. In reconnoitring, more is trusted to the eye.

The scale of the Indian topographical survey is

that of one inch per mile; the scale of half an inch per mile being only resorted to in very densely-wooded or jungly country, containing few inhabitants and little cultivated, or where the climate is so dangerous that it is desirable to accelerate the progress of the survey.

On the scale of one inch per mile the practised draughtsman can survey about five square miles of average country per day. In intricate ground, intersected by ravines or covered by hills of irregular formation, the work proceeds much more slowly; on the other hand, in open and nearly level country, or where the hills have simple outlines, the work will cost less and proceed more rapidly. On the scale of one inch per mile all natural features (such as ravines or water-courses) more than a quarter of a mile in length can be clearly represented. Villages, towns, and cities can be shown, with their principal streets and roads, and the outlines of fortifications. The general figure and extent of cultivated, waste, and forest lands, can be delineated with more or less precision, according to their extent. Irrigated rice-lands should be distinctly indicated, since they generally exhibit the contour of the ground.

The relative heights of hills and depths of valleys should be determined during the course of a topographical survey. These vertical elements of a survey can be ascertained by trigonometrical or by barometrical observations, or by a combination of both methods. "The barometer," says Sir A. Waugh, "is more especially useful for determining the level of low spots from which the principal trigonometrical stations are invisible. In using this instrument, however, in combination with the other operations, the relative differences of heights are to be considered the quantities sought, so that all the results may be referable to the original trigonometrical station. The height above the sea-level of all points coming under any of the following heads are especially to be determined, for the purpose of illustrating the physical relief of the country:

"1. The peaks and highest points of ranges.

"2. All obligatory points required for engineering works, such as roads, drainage, and irrigation, viz.: the highest points or necks of valleys; the lowest depressions or passes in ranges; the junctions of rivers, and *débouchements* of rivers from ranges; the height of inundation-level, at moderate intervals of about three miles apart.

"3. Principal towns or places of note."

Of the various methods employed to indicate the steepness of slope, that of eye-contouring seems alone to merit special comment. In true contouring, regular horizontal lines, at fixed vertical intervals, are traced over a country, and plotted on to the maps. This is an expensive and tedious process, whereas eye-contouring is easy, light, and effective. On this system all that is necessary is that the surveyor should consider what routes persons moving horizontally would pursue. He draws lines on his chart approximating as closely as possible to these imaginary lines. It is evident that when lines are thus drawn for different vertical elevations, the resulting shading will be dark or light, according as the slope is steep or gentle. This method of shading affords scope as well for surveying skill as for draughtsmanship.

(From *Once a Week*, May 1, 1869.)

---

## *A SHIP ATTACKED BY A SWORD-FISH.*

WE have always been puzzled to imagine how the "nine-and-twenty knights of fame," described in the "Lay of the Last Minstrel," managed to "drink the red wine through the helmet barred." But in Nature we meet with animals who seem almost as inconveniently armed as those chosen knights, who

> . . . "quitted not their armor bright,
> Neither by day, nor yet by night."

Among such animals the sword-fish must be recognized as one of the most uncomfortably-armed creatures in existence. The shark has to turn on his back before he can eat, and the attitude scarcely seems suggestive of a comfortable meal. But the sword-fish can hardly even by that arrangement get his awkwardly-projecting snout out of the way. Yet doubtless this feature, which seems so inconvenient, is of great value to Xiphias. In some way as yet unknown it enables him to get his living. Whether he first kills some one of his neighbors with this instrument, and then eats him at his leisure, or whether he plunges it deep into the larger sort of fish, and, attaching himself to them in this way, sucks nutriment from them while they are yet alive, is not known to naturalists. Certainly, he is fond of attacking whales, but this may result not so much from gastronomic tastes as from a natural antipathy—envy, perhaps, at their superior bulk. Unfortunately for himself, Xiphias, though cold-blooded, seems a somewhat warm-tempered animal; and, when he is angered, he makes a bull-like rush upon his foe, without always examining with due care whether he is likely to take any thing by his motion. And when he happens to select for attack a stalwart ship, and to plunge his horny beak through thirteen or fourteen inches of planking, with perhaps a stout copper sheathing outside it, he is apt to find some little difficulty in retreating. The affair usually ends by his leaving his sword embedded in the side of the ship. In fact, no instance has ever been recorded of a sword-fish recovering his weapon (if we may use the expression) after making a lunge of this sort. Last Wednesday the Court of Common Pleas—

rather a strange place, by-the-by, for inquiring into the natural history of fishes—was engaged for several hours in trying to determine under what circumstances a sword-fish might be able to escape scot-free after thrusting his snout into the side of a ship. The gallant ship "Dreadnought," thoroughly repaired, and classed A 1 at Lloyd's, had been insured for £3,000 against all the risks of the seas. She sailed on March 10, 1864, from Colombo, for London. Three days later, the crew, while fishing, hooked a sword-fish. Xiphias, however, broke the line, and a few moments after leaped half out of the water, with the object, it should seem, of taking a look at his persecutor, the "Dreadnought." Probably he satisfied himself that the enemy was some abnormally large Cetacean, which it was his natural duty to attack forthwith. Be this as it may, the attack was made, and at four o'clock the next morning the captain was awakened with the unwelcome intelligence that the ship had sprung a leak. She was taken back to Colombo, and thence to Cochin, where she was hove down. Near the keel was found a round hole, an inch in diameter, running completely through the copper sheathing and planking.

As attacks by sword-fish are included among sea-risks, the insurance company was willing to pay the damages claimed by the owners of the ship, if only it could be proved that the hole had really been made by a sword-fish. No instance had ever been recorded in which a sword-fish had been able to withdraw his sword after attacking a ship. A defence was founded on the possibility that the hole had been made in some other way. Professor Owen and Mr. Frank Buckland gave their evidence; but neither of them could state quite positively whether a sword-fish which had passed its beak through three inches of stout planking could withdraw without the loss of its sword. Mr. Buckland said that fish have no power of "backing," and expressed his belief that he could hold a sword-fish by the beak; but then he admitted that the fish had considerable lateral power, and might so "wriggle its sword out of a hole." And so the insurance company will have to pay nearly six hundred pounds because an ill-tempered fish objected to be hooked, and took its revenge by running full tilt against copper-sheathing and oak-planking.

(From the *Daily News*, December 11, 1868.)

---

## *THE SAFETY-LAMP.*

As the late colliery explosions have attracted a considerable amount of attention to the principle of the safety-lamp, and questions have arisen respecting the extent of the immunity which the action of this lamp secures to the miner, it may be well for us briefly to point out the true qualities of the lamp.

In the Davy lamp a common oil-light is surrounded by a cylinder of wire-gauze. When the air around the lamp is pure the flame burns as usual, and the only effect of the gauze is somewhat to diminish the amount of light given out by the lamp. But so soon as the air becomes loaded with the carburetted hydrogen gas generated in the coal-strata, a change takes place. The flame grows larger and less luminous. The reason of the change is this: The flame is no longer fed by the oxygen of the air, but is surrounded by an atmosphere which is partly inflammable; and the inflammable part of the gas, so fast as it passes within the wire cylinder, is ignited and burns within the gauze. Thus the light now given out by the lamp is no longer that of the comparatively brilliant oil-flame, but is the light resulting from the combustion of carburetted hydrogen, or "fire-damp," as it is called; and every student of chemistry is aware that the flame of this gas has very little illuminating power.

So soon as the miner sees the flame thus enlarged and altered in appearance, he should retire. But it is not true that explosion would necessarily follow if he did not do so. The danger is great, because the flame within the lamp is in direct contact with the gauze, and if there is any defect in the wire-work, the heat may make for itself an opening which—though small—would yet suffice to enable the flame within the lamp to ignite the gas outside. So long, however, as the wire-gauze continues perfect, even though it become red hot, there will be no explosion. No authority is required to establish this point, which has been proved again and again by experiment; but we quote Professor Tyndall's words on the subject to remove some doubts which have been entertained on the matter: "Although a continuous explosive atmosphere," he says, "may extend from the air outside through the meshes of the gauze to the flame within, ignition is not propagated across the gauze. The lamp may be filled with an almost lightless flame; still explosion does not occur. A defect in the gauze, the destruction of the wire at any point by oxidation, hastened by the flame playing against it, would cause explosion;" and so on. It need hardly be said, however, that, imprudent as miners have often been, no miner would remain where his lamp burned with the enlarged flame indicative of the presence of fire-damp. The lamp should also be at once extinguished.

But here we touch on a danger which undoubtedly exists, and—so far as has yet been seen—cannot be guarded against by any amount of caution. Supposing the miner sought to extinguish the lamp by blowing it out, an explosion would almost certainly ensue, since the flame can be forced mechanically through the meshes, though it will not pass through them when it is burning in the ordinary way. Now, of course no miner who had been properly instructed in the use of the safety-lamp would commit such a mistake as this. But it happens, unfortunately, that sometimes the fire-damp itself forces the flame of the lamp through the meshes. The gas frequently issues with great force from cavities in the coal (in which it has been pent up), when the pick of the miner breaks an opening for it. In these circumstances an explosion is inevitable, if the issuing stream of gas happen to be directed full upon the lamp. Fortunately, however, this is a contingency which does not often arise. It is one of those risks of coal-mining which seem absolutely unavoidable by any amount of care or caution. It would be well if it were only such risks as these that the miner had to face.

Another peculiarity sometimes noticed when there is a discharge of fire-damp is worth mentioning. It happens, occasionally, that the light will be put out owing to the absolute exclusion of air from the lamp. This, however, can only happen when the gas issues in so large a volume that the atmosphere of the pit becomes irrespirable.

With the exception of the one risk which we have pointed out above, the Davy lamp may be said to be absolutely safe. It is necessary, however, that caution and intelligence should be exhibited in its use. On this point Professor Tyndall remarks that, unfortunately the requisite intelligence is not often possessed nor the requisite caution exercised by the miner, "and the consequence is that, even with the safety-lamp, explosions still occur." And he suggests that it would be well to exhibit to the miner in a series of experiments the properties of the valuable instrument which has been devised for his security. "Mere advice will not enforce caution," he says; "but let the miner have the physical image of what he is to expect clearly and vividly before his mind, and he will find it a restraining and monitory influence long after the effect of cautioning words has passed away."

A few words on the history of the invention may be acceptable. Early in the present century a series of terrible catastrophes in coal-mines had excited the sympathy of enlightened and humane persons throughout the country. In the year 1813, a society was formed at Sunderland to prevent accidents in coal-mines, or at least to diminish their frequency, and prizes were offered for the discovery of new methods of lighting and ventilating mines. Dr. William Reid Clanny, of Bishopwearmouth, presented to this society a lamp which burned without explosion in an atmosphere heavily loaded with fire-damp; for which invention the Society of Arts awarded him a gold medal. The Rev. Dr. Gray called the attention of Sir Humphry

Davy to the subject, and that eminent chemist visited the coal-mines in 1815 with the object of determining what form of lamp would be best suited to meet the requirements of the coal-miners. He invented two forms of lamp before discovering the principle on which the present safety-lamps are constructed. This principle—the property, namely, that flame will not pass through small apertures—had been, we believe, discovered by Stephenson, the celebrated engineer, some time before; and a somewhat angry controversy took place respecting Davy's claim to the honor of having invented the safety-lamp. It seems admitted, however, by universal consent, that Davy's discovery of the property above referred to was made independently, and also that he was the first to suggest the idea of using wire-gauze in place of perforated tin.

In comparing the present frequency of colliery explosions with what took place before the invention of the safety-lamp, we must take into consideration the enormous increase in the coal-trade since the introduction of steam machinery. The number of miners now engaged in our coal-mines is far in excess of the number employed in the beginning of the present century. Thus accidents in the present day are at once more common on account of the increased rapidity with which the mines are worked, and when they occur there are more sufferers; so that the frequency of colliery explosions in the opening years of the present century, and the number of deaths resulting from them, are in reality much more significant than they seem to be at first sight. But even independently of this consideration, the record of the colliery accidents which took place at that time is sufficiently startling. Seventy-two persons were killed in a colliery at North Biddick at the commencement of the present century. Two explosions in 1805, at Hepburn and Oxclose, left no less that forty-three widows and a hundred and fifty-one children unprovided for. In 1808, ninety persons were killed in a coal-pit at Lumley. On May 24, 1812, ninety-one persons were killed by an explosion at Felling Colliery, near Gateshead. And many more such accidents might readily be enumerated.

(From the *Daily News*, December 4, 1868.)

---

## *THE DUST WE HAVE TO BREATHE.*

A MICROSCOPIST, Mr. Dancer, F. R. A. S., has been examining the dust of our cities. The results are not pleasing. We had always recognized city dust as a nuisance, and had supposed that it derived the peculiar grittiness and flintiness of its structure from the constant macadamizing of city roads. But it now appears that the effects produced by dust, when, as is usual, it finds its way to our eyes, our nostrils, and our throats, are as nothing compared with the mischief it is calculated to produce in a more subtle manner. In every specimen examined by Mr. Dancer, animal life was abundant. But the amount of "molecular activity"—such is the euphuism under which what is exceedingly disagreeable to contemplate is spoken about—is variable according to the height at which the dust is collected. And of all heights which these molecular wretches could select for the display of their activity, the height of five feet is that which has been found to be the favorite. Just at the average height of the foot-passenger's mouth these moving organisms are always waiting to be devoured and to make us ill. And this is not all. As if animal abominations were insufficient, a large proportion of vegetable matter also disports itself in the light dust of our streets. The observations show that in thoroughfares where there are many animals engaged in the traffic, the greater part of the vegetable matter thus floating about "consists of what has passed through the stomachs of animals," or has suffered decomposition in some way or other. This unpleasing matter, like the "molecular activity," floats about at a height of five feet, or thereabouts.

After this, one begins to recognize the manner in which some diseases propagate themselves. What had been mysterious in the history of plagues and pestilences seems to receive at least a partial solution.

Take cholera, for example. It has been shown by the clearest and most positive evidence that this disease is not propagated in any way save one—that is, by the actual swallowing of the cholera-poison. In Professor Thudichum's masterly paper on the subject in the *Monthly Microscopical Journal* it is stated that doctors have inhaled a full breathing from a person in the last stage of this terrible malady without any evil effects. Yet the minutest atom of the cholera-poison received into the stomach will cause an attack of cholera. A small quantity of this matter drying on the floor of the patient's room, and afterward caused to float about in the form of dust, would suffice to prostrate a houseful of people. We can understand, then, how matter might be flung into the streets, and, after drying, its dust be wafted through a whole district, causing the death of hundreds. One of the lessons to be learned from these interesting researches of Mr. Dancer is clearly this, that the watering-cart should be regarded as one of the most important of our hygienic institutions. Supplemented by careful scavenging, it might be effective in dispossessing many a terrible malady which now holds sway from time to time over our towns.

(From the *Daily News*, March 6, 1869.)

## *PHOTOGRAPHIC GHOSTS.*

On the outskirts of the ever-widening circle lighted up by science there is always a border-land wherein superstition holds sway. The arts and sciences may drive away the vulgar hobgoblin of darker days; but they bring with them new sources of illusion. The ghosts of old could only gibber; the spirits of our day can read and write, and play on divers instruments, and quote Shakespeare and Milton. It is not, therefore, altogether surprising to learn that they can take photographs also. You go to have your photograph taken, we will suppose, desiring only to see your own features depicted in the *carte;* and lo! the spirits have been at work, and a photographic phantom makes its appearance beside you. It is true this phantom is of a hazy and dubious aspect—the "dull mechanic ghost" is indistinct, and may be taken for any one. Still, it is not difficult for the eye of fancy to trace in it the lineaments of some departed friend, who, it is to be assumed, has come to be photographed along with you. In fact, photography, according to the spiritualist, resembles what Byron called—

> "The lightning of the mind,
> Which, out of things familiar, undesigned,
> When least we deem of such, calls up to view
> The spectres whom no exorcism can bind."

The phenomena of spiritual photography were first

observed some years since, and a set of carte photographs were sent from America to Dr. Walker, of Edinburgh, in which photographic phantoms were very obviously, however indistinctly, discernible. More recently an English photographer noticed a yet stranger circumstance, though he was too sensible to seek for a supernatural interpretation of it. When he took a photograph with a particular lens, there could be seen not only the usual portrait of the sitter, but at some little distance a faint "double," exactly resembling the principal image. Superstitious minds might find this result even more distressing than the phantom photographic friend. To be visited by the departed through the medium of a lens, is at least not more unpleasing than to hold converse with spirits through an ordinary "rapping" medium. But the appearance of a "double" or "fetch," has ever been held by the learned in ghostly lore to signify approaching death.

Fortunately, both one and the other appearance can be very easily accounted for without calling in the aid of the supernatural. At a recent meeting of the Photographical Society it was shown that an image may often be so deeply impressed on the glass that the subsequent cleaning of the plate, even with strong acids, will not completely remove the picture. When the plate is used for receiving another picture, the original image makes its reappearance, and as it is too faint to be recognizable, a highly-susceptible imagination may readily transform it into the image of a departed friend. The "double" is generated by the well-known property of double refraction, obtained by a lens under certain circumstances of unequal pressure, or sometimes by inequalities in the process of annealing. So vanish two ghosts which might have been more or less troublesome to those who are ready to see the supernatural in commonplace phenomena. Will the time ever come when no more such phantoms will remain to be exorcised?

(From the *Daily News*, March 2, 1869.)

---

## *THE OXFORD AND CAMBRIDGE ROWING STYLES.*

Whatever opinion we may have of the result of the approaching contest (1869), there can be no doubt that this year, as in former years, there is a striking dissimilarity between the rowing styles of the dark-blue and the light-blue oarsmen. This dissimilarity makes itself obvious whether we compare the two boats as seen from the side, or when the line of sight is directed along the length of either. Perhaps it is in the latter aspect that an unpractised eye will most readily detect the difference we are speaking of. Watch the Cambridge boat approaching you from some distance, or receding, and you will notice in the rise and fall of the oars, as so seen, the following peculiarities—a long stay of the oar in the water, a quick rise from and return to the water, the oars remaining out of the water for the briefest possible interval of time. In the case of the Oxford boat quite a different appearance is presented—there is a short stay in the water, a sharp rise from and return to it, and between these the oars appear to hang over the water for a perceptible interval. It is, however, when the boats are seen from the side that the meaning of these peculiarities is detected, and also that the fundamental distinction between the two styles is made apparent to the experienced eye. In the Cambridge boat we recognize the long stroke and "lightning feather" inculcated in the old treatise on rowing: in the Oxford boat we see these conditions reversed, and in their place the "waiting feather" and lightning stroke. By the "waiting feather," we do not refer to what is commonly understood by slow feathering, but to a momentary pause (scarcely to be detected when the crew is rowing hard) before the simultaneous dash of the oars upon the first grip of the stroke.* And observing more closely—which, by-the-way, is no easy matter—as either boat dashes swiftly past, we detect distinctive peculiarities of "work" by which the two styles are severally arrived at. In the Cambridge crew we see the first part of

* The grip is never properly caught without the pause; but any thing beyond a momentary pause is a bad fault in style.

the stroke done with the shoulders—precisely according to the old-fashioned models—the arms straight until the body has fallen back to an almost upright position; then comes the sharp drop back of the shoulders beyond the perpendicular, the arms simultaneously doing their work, so that as the swing back is finished, the backs of the hands just touch the ribs in feathering. All these things are quite in accordance with what used to be considered the perfection of rowing; and, indeed, this style of rowing has some important good qualities and a very handsome appearance. The lightning feather, also, which follows the long sweeping stroke, is theoretically perfect. Now, in the case of the Oxford crew, we observe a style which at first sight seems less excellent. As soon as the oars are dashed down and catch their first hold of the water, the arms as well as the shoulders of each oarsman are at work.* The result is, that when the

* I write this with full knowledge that many Oxford men deny the fact. I have rowed behind Cambridge, Oxford, and London strokes, and have several times taken the place (number 2 thwart) of a London waterman in a four ("stroke" by John Mackinney) training for the Thames Regatta. So that I have had ample opportunities for comparing different rowing styles; and I am satisfied that the main defect of the real Cambridge style was (and perhaps is) an *exaggeration* of the sound rule that a boat should be propelled rather by the body than by the arms. The very swing in a Cambridge boat shows that this must be so. On the other hand, the Thames watermen do too much arm-work; and hence seem to double a little over their oars. I once rowed with some Cambridge friends from London nearly to Oxford and back, taking a Thames waterman as "help." We set him, at first, for our strokesman, but presently made him row *bow*, for we could none of us stand his gripping, arm-working style.

back has reached an upright position, the arms have already reached the chest, and the stroke is finished. Thus the Oxford stroke takes a perceptibly shorter time than the Cambridge stroke; it is also, necessarily, somewhat shorter in the water. One would, therefore, say it must be less effective. Especially would an unpractised observer form this opinion, because the Oxford stroke seems to be much shorter in range than it is in reality. *There* we have the secret of its efficiency. It is actually nearly as long as the Cambridge stroke, but is taken in a perceptibly shorter time. What does this mean but that the oar is taken much more sharply, and, therefore, much more effectively, through the water?

Much more effectively so far as the actual conditions of the contest are concerned. The modern racing outrigger requires a sharp impulse, because it will take almost any speed we can apply to it. It will also retain that speed between the strokes, a consideration of great importance. The old-fashioned racing-eights required to be continually under propulsion. The lightning-feather was a necessity in their case, for between every stroke the boat would lag terribly with a slow-feathering crew. We do not say, of course, that the speed of a light outrigged craft does not diminish between the strokes. Any one who has watched a closely-contested bumping-race, and noticed the way in which the sharply-cut bow of the pursuing boat draws up to the rudder of the other as by a succession of impulses, although either boat seen alone would seem to sweep on with almost uniform speed, will know that the motion of the lightest boat is not strictly uniform. But there is an immense difference between the almost imperceptible loss of way of a modern eight and the dead "lag" in the old-fashioned craft. And hence we get the following important consideration: Whereas with the old boats it was useless for a crew to attempt to give a very quick motion to their boat by a sharp, sudden "lift," this plan is calculated to be, of all others, the most effective with the modern racing-eight.

It may seem, at first sight, that, after all, the result of the Cambridge style should be as effective as that of the other. If arms and shoulders do their work in both crews with equal energy—which we may assume to be the case—and if the number of strokes per minute is equal, the actual propulsive energy ought to be equal likewise. A little consideration will show that this is a fallacy. If two men pull at a weight together they will move it farther with a given expenditure of energy than if first one and then the other apply his strength to the work. And, what is more to the purpose, they will be able to move it faster. So shoulders and arms working simultaneously will give a greater propulsive power than when working separately, even though in the latter case each works with its fullest energy. And not only so, but by the simultaneous use of arms and shoulders, that sharpness of motion can alone be given which is essential to the propulsion of a modern racing-boat.

We have said that the two crews are severally rowing in the style which has lately been peculiar to their respective universities. But the Cambridge crew is rowing in that form of the Cambridge style which brings it nearest to the requirements of modern racing. The faults of the style are subdued, so to speak, and its best qualities brought out effectually. In one or two of the long series of defeats lately sustained by Cambridge the reverse has been the case. At present, too, there is a certain roughness about the Oxford crew which encourages the hopes of the light-blue supporters. But it must be admitted that this roughness is rather apparent than real, great as it seems, and it will doubtless disappear before the day of encounter. We venture to predict that the "time" of the approaching race, taken in conjunction with the state of the tide, will show the present crews to be at least equal to the average.*

* The race (that of 1869) was one of the best ever rowed, and the time of the winners (Oxford) better than in any former race.

(From the *Daily News*, April, 1869.)

## *BETTING ON HORSE-RACES: OR, THE STATE OF THE ODDS.*

There appears every day in the newspapers an account of the betting on the principal forthcoming races. The betting on such races as the Two Thousand Guineas, the Derby, and the Oaks, often begins more than a year before the races are run; and during the interval, the odds laid against the different horses engaged in them vary repeatedly, in accordance with the reported progress of the animals in their training, or with what is learned respecting the intentions of their owners. Many who do not bet themselves, find an interest in watching the varying fortunes of the horses which are held by the initiated to be leading favorites, or to fall into the second rank, or merely to have an outside chance of success. It is amusing to notice, too, how frequently the final state of the odds is falsified by the event; how some "rank outsider" will run into the first place, while the leading favorites are not even "placed."

It is in reality a simple matter to understand the betting on races (or contests of any kind), yet it is astonishing how seldom those who do not actually bet upon races have any inkling of the meaning of those mysterious columns which indicate the opinion of the betting world respecting the probable results of approaching contests, equine or otherwise.

Let us take a few simple cases of "odds," to begin with; and, having mastered the elements of our subject, proceed to see how cases of greater complexity are to be dealt with.

Suppose the newspapers inform us that the betting is 2 to 1 against a certain horse for such and such a race, what inference are we to deduce? To learn this, let us conceive a case in which the *true* odds against a certain event are as 2 to 1. Suppose there are three balls in a bag, one being white, the others black. Then, if we draw a ball at random, it is clear that we are twice as likely to draw a black as to draw a white ball. This is technically expressed by saying that the odds are 2 to 1 *against* drawing a white ball; or 2 to 1 *on* (that, is in favor of) drawing a black ball. This being understood, it follows that, when the odds are said to be 2 to 1 against a certain horse, we are to infer that, in the opinion of those who have studied the performance of the horse, and compared it with that of the other horses engaged in the race, his chance of winning is equivalent to the chance of drawing one particular ball out of a bag of three balls.

Observe how this result is obtained: the odds are 2 to 1, and the chance of the horse is as that of drawing one ball out of a bag of three—three being the sum of the two numbers 2 and 1. This is the method followed in all such cases. Thus, if the odds against a horse are 7 to 1, we infer that the *cognoscenti* consider his chance equal to that of drawing one particular ball out of a bag of *eight*.

A similar treatment applies when the odds are not given as so many to *one*. Thus, if the odds against a horse are as 5 to 2, we infer that the horse's chance is equal to that of drawing a white ball out of a bag containing five black and two white balls—or seven in all.

We must notice also that the number of balls may be increased to any extent, provided the proportion between the total number and the number of a specified color remains unchanged. Thus, if the odds are 5 to 1 against a horse, his chance is assumed to be equivalent to that of drawing *one* white ball out of a bag containing six balls, only one of which is white; *or* to that of drawing a white ball out of a bag containing sixty balls, of which ten are white—and so on. This is a very important principle, as we shall now see.

Suppose there are two horses (among others) engaged in a race, and that the odds are 2 to 1 against one, and 4 to 1 against the other—what are the odds that one of the two horses will win the race? This case will doubtless remind our readers of an amusing sketch by Leech, called—if we remember rightly—"Signs of the Commission." Three or four undergraduates are at a "wine," discussing matters equine.

One propounds to his neighbor the following question: "I say, Charley, if the odds are 2 to 1 against *Rataplan*, and 4 to 1 against *Quick March*, what's the betting about the pair?"—"Don't know, I'm sure," replies Charley; "but I'll give you 6 to 1 against them." The absurdity of the reply is, of course, very obvious; we see at once that the odds cannot be heavier against a pair of horses than against either singly. Still there are many who would not find it easy to give a correct reply to the question. What has been said above, however, will enable us at once to determine the just odds in this or any similar case. Thus—the odds against one horse being 2 to 1, his chance of winning is equal to that of drawing one white ball out of a bag of *three*, one only of which is white. In like manner, the chance of the second horse is equal to that of drawing one white ball out of a bag of *five*, one only of which is white. Now we have to find a number which is a multiple of both the numbers three and five. Fifteen is such a number. The chance of the first horse, modified according to the principle explained above, is equal to that of drawing a white ball out of a bag of fifteen of which *five* are white. In like manner, the chance of the second is equal to that of drawing a white ball out of a bag of fifteen of which *three* are white. Therefore, the chance that *one of the two* will win is equal to that of drawing a white ball out of a bag of fifteen balls of which *eight* (five added to three) are white. There remain seven black balls, and therefore the odds are 8 to 7 *on* the pair.

To impress the method of treating such cases on the mind of the reader, we take the betting about three horses—say 3 to 1, 7 to 2, and 9 to 1, *against* the three horses respectively. Then their respective chances are equal to the chance of drawing (1) one white ball out of *four*, one only of which is white; (2) a white ball out of *nine*, of which two only are white; and (3) one white ball out of *ten*, one only of which is white. The least number which contains four, nine, and ten, is 180; and the above chances, modified according to the principle explained above, become equal to the chance of drawing a white ball out of a bag containing 180 balls, when 45, 40, and 18 (respectively) are white. Therefore, the chance that one of the three will win is equal to that of drawing a white ball out of a bag containing 180 balls, of which 103 (the sum of 45, 40, and 18) are white. Therefore, the odds are 103 to 77 *on* the three.

One does not hear in practice of such odds as 103 to 77. But betting-men (whether or not they apply just principles of computation to such questions, is unknown to us) manage to run very near the truth. For instance, in such a case as the above, the odds on the three would probably be given as 4 to 3—that is, instead of 103 to 77, or, which is the same thing, 412 to 308, the published odds would be 412 to 309.

And here a certain nicety in betting has to be mentioned. In running the eye down the list of odds, one will often meet such expressions as 10 to 1 against such a horse *offered*, or 10 to 1 *wanted*. Now, the odds of 10 to 1 *taken* may be understood to imply that the horse's chance is equivalent to that of drawing a certain ball out of a bag of eleven. But if the odds are offered and not taken, we cannot infer this. The offering of the odds implies that the horse's chance is *not better* than that above mentioned, but the fact that they are not taken implies that the horse's chance is *not so good*. If no higher odds are offered against the horse, we may infer that his chance is *very little worse* than that mentioned above. Similarly, if the odds of 10 to 1 are *asked for*, we infer that the horse's chance is *not worse* than that of drawing one ball out of eleven; if the odds are not obtained, we infer that his chance is *better;* and if no lower odds are asked for, we infer that his chance is *very little better*.

Thus, there might be *three* horses (A, B, and C) against whom the nominal odds were 10 to 1, and yet these horses might not be equally good favorites, because the odds might not be taken, or might be asked for in vain. We might accordingly find three such horses arranged thus:

| | Odds. |
|---|---|
| A . . . | 10 to 1 (wanted). |
| B . . . | 10 to 1 (taken). |
| C . . . | 10 to 1 (offered). |

Or these different stages might mark the upward or downward progress of the same horse in the betting. In fact, there are yet more delicate gradations, marked by such expressions respecting certain odds, as —*offered freely*, *offered*, *offered and taken* (meaning that some offers only have been accepted), *taken*, *taken and wanted*, *wanted*, and so on.

As an illustration of some of the principles we have been considering, let us take from the day's paper* the state of the odds respecting the "Two Thousand Guineas." It is presented in the following form:

TWO THOUSAND GUINEAS.

7 to 2 against *Rosicrucian* (off.).
6 to 1 against *Pace* (off.; 7 to 1 w.).
10 to 1 against *Green Sleeve* (off.).
100 to 7 against *Blue Gown* (off.).
180 to 80 against Sir J. Hawley's lot (t.).

This table is interpreted thus: bettors are willing to lay the same odds against *Rosicrucian* as would be the true mathematical odds against drawing a white ball out of a bag containing two white and seven black balls; but no one is willing to back the horse at this rate. On the other hand, higher odds are not offered against him. Hence it is presumable that his chance is somewhat less than that above indicated. Again, bettors are willing to lay the same odds against *Pace* as might fairly be laid against drawing one white ball

* This article was written early in March, 1868.

out of a bag of seven, one only of which is white; but backers of the horse consider that they ought to get the same odds as might be fairly laid against drawing the white ball when an additional black ball had been put into the bag. As respects *Green Sleeve* and *Blue Gown*, bettors are willing to lay the odds which there would be, respectively, against drawing a white ball out of a bag containing—(1) eleven balls, one only of which is white, and (2) one hundred and seven balls, seven only of which are white. Now, the three horses *Rosicrucian*, *Green Sleeve*, and *Blue Gown*, all belong to Sir Joseph Hawley, so that the odds about the three are referred to in the last statement of the list just given. And since none of the offers against the three horses have been taken, we may expect the odds actually taken about "Sir Joseph Hawley's lot" to be more favorable than those obtained by summing up the three former in the manner we have already examined. It will be found that the resulting odds (offered) against Sir J. Hawley's lot—estimated in this way—should be, as nearly as possible, 132 to 80. We find, however, that the odds *taken* are 180 to 80. Hence, we learn that the offers against some or all of the three horses are considerably short of what backers require; or else, that some person has been induced to offer far heavier odds against Sir J. Hawley's lot than are justified by the fair odds against his horses, severally.

We have heard it asked why a horse is said to be a favorite, though the odds may be against him. This is very easily explained. Let us take as an illustration the case of a race in which four horses are engaged to run. If all these horses had an equal chance of winning, it is very clear that the case would correspond to that of a bag containing four balls of different colors; since, in this case, we should have an equal chance of drawing a ball of any assigned color. Now, the odds against drawing a particular ball would clearly be 3 to 1. This, then, should be the betting against each of the three horses. If any one of the horses has less odds offered against him, he is *a favorite*. There may be more than one of the four horses thus distinguished; and in that case, the horse against which the least odds are offered is *the first favorite*. Let us suppose there are two favorites, and that the odds against the leading favorite are 3 to 2, those against the other 2 to 1, and those against the best non-favorite 4 to 1; and let us compare the chance of the four horses. We have not named any odds against the fourth, because, if the odds against all the horses but one are given, the just odds against that one are determinable, as we shall see immediately. The chance of the leading favorite corresponds to the chance of drawing a ball out of a bag in which are three black and two white balls, *five* in all; that of the next to the chance of drawing a ball out of a bag in which are two black and one white ball, *three* in all; that of the third, to the chance of drawing a ball out of a bag in which are four black balls and one white one, *five* in all. We take, then, the least number containing both five and three—that is, *fifteen;* and then the number of white balls, corresponding to the chances of the three horses, are respectively six, five, and three, or fourteen in all; leaving only *one* to represent the chance of the fourth horse (against which the odds are, therefore, 14 to 1). Hence the chances of the four horses are respectively as the numbers *six*, *five*, *three*, and *one*.

We have spoken above of the published odds. The statements made in the daily papers commonly refer to wagers actually made, and therefore the uninitiated might suppose that every one who tried would be able to obtain the same odds. This is not the case. The wagers which are laid between practised betting-men afford very little indication of the prices which would be forced (so to speak) upon an inexperienced bettor. Book-makers—that is, men who make a series of bets upon several or all of the horses engaged in a race—naturally seek to give less favorable terms than the known chances of the different horses engaged would suffice to warrant. As they cannot offer such terms to the initiated, they offer them—and in general successfully—to the inexperienced.

It is often said that a man may so lay his wagers about a race as to make sure of gaining money whichever horse wins the race. This is not strictly the case. It is of course possible to make sure of winning if the bettor can only get persons to lay or take *the odds he requires to the amount he requires.* But this is precisely the problem which would remain insoluble if all bettors were equally experienced.

Suppose, for instance, that there are three horses engaged in a race with equal chances of success. It is readily shown that the odds are 2 to 1 against each. But if a bettor can get a person to take even betting against the first horse (A), a second person to do the like about the second horse (B), and a third to do the like about the third horse (C), and if all these bets are made to the same amount—say, £1,000—then, inasmuch as only one horse can win, the bettor loses £1,000 on that horse (say A), and gains the same sum on each of the two horses B and C. Thus, on the whole, he gains £1,000, the sum laid out against each horse.

If the layer of the odds had laid the true odds to the same amount on each horse, he would neither have gained nor lost. Suppose, for instance, that he laid £1,000 to £500 against each horse, and A won; then he would have to pay £1,000 to the backer of A, and to receive £500 from each of the backers of B and C. In like manner, a person who had backed each horse to the same extent would neither lose nor gain by the event. Nor would a backer or layer who had wagered *different* sums *necessarily* gain or lose by the race; he would gain or lose *according to the event.* This will at once be seen, on trial:

Let us next take the case of horses with unequal prospects of success—for instance, take the case of the four horses considered above, against which the odds were respectively 3 to 2, 2 to 1, 4 to 1, and 14 to 1. Here, suppose the same sum laid against each, and for convenience let this sum be £84 (because 84 contains the numbers 3, 2, 4, and 14). The layer of the odds wagers £84 to £56 against the leading favorite, £84 to £42 against the second horse, £84 to £21 against the third, and £84 to £6 against the fourth. Whichever horse wins, the layer has to pay £84; but if the favorite wins, he receives only £42 on one horse, £21 on another, and £6 on the third—that is £69 in all, so that he loses £15; if the second horse wins, he has to receive £56, £21, and £6—or £83 in all, so that he loses £1; if the third horse wins, he receives £56, £42, and £6—or £104 in all, and thus gains £20; and, lastly, if the fourth horse wins, he has to receive £56, £42, and £21—or 119 in all, so that he gains £35. He clearly risks much less than he has a chance (however small) of gaining. It is also clear that in all such cases the worst event for the layer of the odds is, that the favorite should win. Accordingly, as professional book-makers are nearly always layers of odds, one often finds the success of a favorite spoken of in the papers as a "great blow for the book-makers," while the success of a rank outsider will be described as "a misfortune to backers."

But there is another circumstance which tends to make the success of a favorite a blow to layers of the odds, and *vice versâ.* In the case we have supposed, the money actually pending about the four horses (that is, the sum of the amounts laid *for* and *against* them) was £140 as respects the favorite, £126 as respects the second, £105 as respects the third, and £90 as respects the fourth. But, as a matter of fact, the amounts pending about the favorites bear always a much greater proportion than the above to the amounts pending about outsiders. It is easy to see the effect of this. Suppose, for instance, that instead of the sums £84 to £56, £84 to £42, £84 to £21, and £84 to £6, a book-maker had laid £8,400 to £5,600, £840 to £420, £84 to £21, and £14 to £1, respectively—then it will easily be seen that he would lose £7,958 by the success of the favorite; whereas he would gain £4,782 by the success of the second horse, £5,937 by that of the third, and £6,027 by that of the fourth. We have taken this as an extreme case; as a general rule, there is not so great a disparity as has been here assumed between the sums pending on favorites and outsiders.

Finally, it may be asked whether, in the case of horses having unequal chances, it is possible that wagers can be so proportioned (just odds being given and taken) that, as in the former case, a person backing, or laying against, all the four shall neither gain nor lose. It is so. All that is necessary is, that the sum actually pending about each horse shall be the same. Thus, in the preceding case, if the wagers £9 to £6, £10 to £5, £12 to £3, and £14 to £1, are either laid or taken by the same person, he will neither gain nor lose by the event, whatever it may be. And, therefore, if unfair odds are laid or taken about all the horses, in such a manner that the amounts pending on the several horses are equal (or nearly so), the unfair bettor must win by the result. Say, for instance, that instead of the above odds, he lays £8 to £6, £9 to £5, £11 to £3, and £13 to £1 against the four horses respectively; it will be found that he *must* win £1. Or if he *takes* the odds £18 to £11, £20 to £9, £24 to £5, and £28 to £1 (the just odds being £18 to £12, £20 to £10, £24 to £6, £28 to £2 respectively), he will win £1 by the race. So that, by giving or taking such odds to a sufficiently great amount, a bettor would be certain of pocketing a large sum, whatever the event of a given race might be.

In every instance, a. man who bets on a race *must risk his money*, unless he can succeed in taking unfair advantages over those with whom he bets. Our readers will conceive how small must be the chance that an unpractised bettor will gain any thing but dearly-bought experience by speculating on horse-races. We would recommend those who are tempted to hold another opinion to follow the plan suggested by Thackeray in a similar case—to take *a good look* at professional and practised betting-men, and to decide "which of those men they are most likely to get the better of" in turf transactions.

(From *Chambers's Journal*, July, 1869.)

---

## *SQUARING THE CIRCLE.*

THERE must be a singular charm about insoluble problems, since there are never wanting persons who are willing to attack them. We doubt not that at this moment there are persons who are devoting their energies to Squaring the Circle, in the full belief that important advantages would accrue to science—and possibly a considerable pecuniary profit to themselves—if they could succeed in solving it. Quite recently, applications have been made to the Paris Académy of Sciences, to ascertain what was the amount which that body was authorized to pay over to any one who should square the circle. So seriously, indeed, was the secretary annoyed by applications of this sort, that it was found necessary to announce in the daily journals that not only was the Academy not authorized to pay any sum at all, but that it had determined never to give the least attention to those who fancied they had mastered the famous problem.

It is a singular circumstance that people have even attacked the problem without knowing exactly what its nature is. One ingenious workman, to whom the difficulty had been propounded, actually set to work to invent an arrangement for measuring the circumference of the circle; and was perfectly satisfied that he had thus solved a problem which had mastered all the mathematicians of ancient and modern times. That we may not fall into a similar error, let us clearly understand what it is that is required for the solution of the problem of "squaring the circle."

To begin with, we must note that the term "squaring the circle" is rather a misnomer; because the true problem to be solved is the determination of the length of a circle's circumference when the diameter is known. Of course, the solution of this problem, or, as it is termed, the *rectification* of the circle, involves the solution of the other, or the *quadrature* of the circle. But it is well to keep the simpler issue before us.

Many have supposed that there exists some exact relation between the circumference and the diameter of the circle, and that the problem to be solved is the determination of this relation. Suppose, for example, that the approximate relation discovered by Archimedes (who found, that if a circle's diameter is represented by *seven*, the circumference may be almost exactly represented by *twenty-two*) were strictly correct, and that Archimedes had proved it to be so: then, according to this view, he would have solved the great problem; and it is to determine a relation of some such sort that many persons have set themselves. Now, undoubtedly, if any relation of this sort could be established, the problem would be solved; but, as a matter of fact, no such relation exists, and the solution of the problem does not require that there should be any relation of the sort. For example, we do not look on the determination of the diagonal of a square (whose side is known) as an insoluble, or as otherwise than a very simple problem. Yet in this case no exact relation exists. We cannot possibly express both the side and the diagonal of a square in whole numbers, no matter what unit of measurement we adopt: or, to put the matter in another way, we cannot possibly divide both the side and the diagonal into equal parts (which shall be the same along each), no matter how small we take the parts. If we divide the side into 1,000 parts, there will be 1,414 such parts, *and a piece over*, in the diagonal; if we divide the side into 10,000 parts, there will be 14,142, and still a little piece over, in the diagonal; and so on for ever. Similarly, the mere fact that no exact relation exists between the diameter and the circumference of a circle is no bar whatever to the solution of the great problem.

Before leaving this part of the subject, however, we may mention a relation which is very easily remembered, and is very nearly exact—much more so, at any rate, than that of Archimedes. Write down the numbers 113355, that is, the first three odd numbers each repeated twice over. Then separate the six numbers into two sets of three, thus: 113)355, and proceed with the division thus indicated. The result, 3.1415929 . . . . , expresses the circumference of a circle whose diameter is 1, correctly to the sixth decimal place, the true relation being 3.14159265 . . . .

Again, many people imagine that mathematicians are still in a state of uncertainty as to the relation which exists between the circumference and the diameter of the circle. If this were so, scientific societies might well hold out a reward to any one who could enlighten them; for the determination of this relation (with satisfactory exactitude) may be held to lie at the foundation of the whole of our modern system of mathematics. We need hardly say that no doubt whatever rests on the matter. A hundred different methods are known to mathematicians by which the circumference may be calculated from the diameter with any required degree of exactness. Here is a simple one, for example: Take any number of the fractions formed by putting *one* as a numerator over the successive odd numbers. Add together the alternate ones beginning with the first, which, of course, is unity. Add together the remainder. Subtract the second sum from the first. The remainder will express the circumference (the diameter being taken as unity) to any required degree of exactness. We have merely to take enough fractions. The process would, of course, be a very laborious one, if great exactness were required, and as a matter of fact, mathematicians have made use of much more convenient methods for determining the required relation; but the method is strictly exact.

The largest circle we have much to do with in scientific questions is the earth's equator. As a matter of curiosity, we may inquire what the circumference of the earth's orbit is; but as we are far from being sure of the exact length of the radius of that orbit (that is, of the earth's distance from the sun), it is clear that we do not need a very exact relation between the circumference and the diameter in dealing with that enormous circle. Confining ourselves, therefore, to the circle of the earth's equator, let us see what exactness we seem to require. We will suppose for a moment that it is possible to measure round the earth's equator without losing count of a single yard, and that we want to gather from our estimate what the diameter of this great circle may be. This seems, indeed, the only use to which, in this case, we can put our knowledge of the relation we are dealing with. We have then a circle some twenty-five thousand miles round, and each mile contains one thousand seven hundred and sixty yards; or, in all, there are some forty-four million yards in the circumference, and therefore (roughly) some fourteen million yards in the diameter of this great circle. Hence, if our relation is correct within a fourteen-millionth part of the diameter, or forty-four-millionth part of the circumference, we are safe from any error exceeding a yard. All we want, then, is that the number expressing the circumference (the diameter being unity) should be true to the eighth decimal place, as quoted above.

But, as we have said, mathematicians have not been content with a computation of this sort. They have calculated the number not to the *eighth*, but to the *six hundred and twentieth* decimal place. Now, if we remember that each new decimal makes the result ten times more exact, we shall begin to see what a waste of time there has been in this tremendous calculation. We all remember the story of the horse which had twenty-four nails in its shoes, and was valued at the sum obtained by adding together a farthing for the first nail, a halfpenny for the next, a penny for the next, and so on; doubling twenty-four times. The result was counted by thousands of pounds. The old miser who paid at a similar rate for a grave eighteen feet deep (doubling for each foot), killed himself when he heard the total. But now consider the effect of multiplying by ten, six hundred and twenty times. A fraction, with that enormous number for denominator, and unity for numerator, expresses the minuteness of the error which would result if the "long value" of the circumference were made use of. Let an illustration present the meaning of this:—

It has been estimated that light, which could eight times circle the earth in a second, takes 50,000 years in reaching us from the faintest star seen in Lord Rosse's giant reflector. Suppose we knew the exact length of the tremendous line which extends from the earth to such a star, and wanted, for some inconceivable purpose, to know the length of the circumference of a circle of which that line was the radius. The value deduced from the above-mentioned calculation of the relation between the circumference and the the diameter would differ from the truth by a length which would be imperceptible under the most powerful microscope ever yet constructed. Nay, the radius we have conceived, enormous as it is, might be increased a million-fold, or a million times a million-fold, with the same result. And the area of the circle formed with this increased radius would be determinable with so much accuracy, that the error, if presented in the form of a minute square, would be utterly imperceptible under a microscope a million times more powerful than the best ever yet constructed by man.

Not only has the length of the circumference been calculated once in this unnecessarily exact manner, but a second calculator has gone over the work independently. The two results are of course identical figure for figure.

It will be asked then, what *is* the problem about which so great a work has been made? The problem is, in fact, utterly insignificant; its only interest lies in the fact that it is insoluble—a property which it shares along with many other problems, as the trisection of an angle, the duplication of a cube, and so on.

The problem is simply this: *Having given the diameter of a circle, to determine, by a geometrical construction, in which only straight lines and circles shall be made use of, the side of a square equal in area to the circle.* As we have said, the problem is solved, if, by a construction of the kind described, we can determine the length of the circumference; because then the rectangle under half this length and the radius is equal in area to the circle, and it is a simple problem to describe a square equal to a given rectangle.

To illustrate the kind of construction required, we give an approximate solution which is remarkably simple, and, so far as we are aware, not generally known. Describe a square about the given circle, touching it at the ends of two diameters, AOB, COD, at right angles to each other, and join CA; let COAE be one of the quarters of the circumscribing square, and from E draw EG, cutting off from AO a fourth part AG of its length, and from AC the portion AH. Then three sides of the circumscribing square together with AH are very nearly equal to the circumference of the circle. The difference is so small, that in a circle two feet in diameter, it would be less than the two-hundredth part of an inch. If this construction were exact, the great problem would have been solved.

One point, however, must be noted: the circle is of all curved lines the easiest to draw by mechanical means. But there are others which can be so drawn. And, if such curves as these be admitted as available, the problem of the quadrature of the circle can be readily solved. There is a curve, for instance, invented by Dinostratus which can readily be described mechanically, and has been called the quadratrix of Dinostratus, because it has the property of thus solving the problem we are dealing with.

As such curves can be described with quite as much accuracy as the circle—for, be it remembered, an absolutely perfect circle has never yet been drawn—we see that it is only the limitations which geometers have themselves invented that give this problem its difficulty. Its solution has, as we have said, no value; and no mathematician would ever think of wasting a moment over the problem—for this reason, simply, that it has long since been demonstrated to be insoluble by simple geometrical methods. So that, when a man says he has squared the circle (and many will say so, if one will only give them a hearing), he shows that either he wholly misunderstands the nature of the problem, or that his ignorance of mathematics has led him to mistake a faulty for a true solution.

(From *Chambers's Journal*, January 16, 1869.)

---

## *THE NEW THEORY OF ACHILLES'S SHIELD.*

A distinguished classical authority has remarked that the description of Achilles's shield occupies an anomalous position in Homer's "Iliad." On the one hand, it is easy to show that the poem—for the description may be looked on as a complete poem—is out of place in the "Iliad;" on the other, it is no less easy to show that Homer has carefully led up to the description of the shield by a series of introductory events.

I propose to examine, briefly, the evidence on each of these points, and then to exhibit a theory respecting the shield which may appear *bizarre* enough on a first view, but which seems to me to be supported by satisfactory evidence.

An argument commonly urged against the genuineness of the "Shield of Achilles" is founded on the length and labored character of the description. Even Grote, whose theory is that Homer's original poem was not an *Iliad*, but an *Achilleis*, has admitted the force of this argument. He finds clear evidence that from Book II. to Book XX., Homer has been husbanding his resources for the more effective description of the final conflict. He therefore concedes the possibility that the "Shield of Achilles" may be an interpolation —perhaps the work of another hand.

It appears to me, however, that the mere length of the description is no argument against the genuineness of the passage. Events have, indeed, been hastening to a crisis up to the end of Book XVII., and the action is checked in a marked manner by the 'Oplopœia" in Book XVIII. Yet it is quite in Homer's manner to introduce, between two series of important events, an interval of comparative inaction, or at least of events wholly different in character from those of either series. We have a marked instance of this in Books IX. and X. Here the appeal to Achilles and the night-adventure of Diomed and Ulysses are interposed between the first victory of the Trojans and the great struggle in which Patroclus is slain, and Agamemnon, Ulysses, Diomed, Machaon, and Eurypylus wounded.* In fact

* Another well-known instance, where "Patroclus sent in hot haste for news by a man of the most fiery impatience, is button-held by Nestor, and though he has no time to sit down, yet is obliged to endure a speech of 152 lines," is accounted for by Gladstone in a different manner.

one cannot doubt that in such an arrangement Homer exhibits admirable taste and judgment. The contrast between action and inaction, or between the confused tumult of a heady conflict and the subtle advance of the two Greek heroes, is conceived in the true poetic spirit. The dignity and importance of the action, and the interest of the interposed events, are alike enhanced. Indeed, there is scarcely a noted author whose works do not afford instances of corresponding contrasts. How skilfully, for example, has Shakespeare interposed the "bald, disjointed chat" of the sleepy porter between the conscience-wrought horror of Duncan's murderers and the "horror, horror, horror" which "tongue nor heart could not conceive nor name" of his faithful followers. Nor will the reader need to be reminded of the frequent and effective use by Dickens of the contrast between the humorous and the pathetic.

The labored character of the description of the shield is an argument—though not, perhaps, a very striking one—of the independent origin of the poem.

But the arguments on which I am disposed to lay most stress lie nearer the surface.

Scarcely any one, I think, can have read the description of the shield without a feeling of wonder that Homer should describe the shield of a mortal hero as adorned with so many and such important objects. We find the sun and moon, the constellations, the waves of ocean, and a variety of other objects, better suited to adorn the temple of a great deity than the shield of a warrior, however noble and heroic. The objects depicted even on the Ægis of Zeus are much less important. There is certainly no trace in the "Iliad" of a wish on Homer's part to raise the dignity of mortal heroes at the expense of Zeus, yet the Ægis is thus succinctly described:

> "Fringed round with ever-fighting snakes, though it was drawn
> to life
> The miseries and deaths of fight; in it frowned bloody Strife,
> In it shone sacred Fortitude, in it fell Pursuit flew,
> In it the monster Gorgon's head, in which held out to view
> Were all the dire ostents of Jove."—*Chapman's* Translation.

Five lines here, as in the original, suffice for the description of Jove's Ægis, while one hundred and thirty lines are employed in the description of the celestial and terrestrial objects depicted on the shield of Achilles.

Another circumstance attracts notice in the description of Achilles's armor—the disproportionate importance attached to the shield. Undoubtedly, the shield was that portion of a hero's armor which admitted of the freest application of artistic skill. Yet this consideration is not sufficient to account for the fact that, while so many lines are given to the shield, the helmet, corselet, and greaves, are disposed of in four.

But the argument on which I am inclined to lay most stress is the occurrence *elsewhere* of a description which is undoubtedly only another version of the "Shield of Achilles." The "Shield of Hercules" occurs in a poem ascribed to Hesiod. But whatever opinion may be formed respecting the authorship of the description, there can be no doubt that it is not Hesiod's work. It exhibits no trace of his dry, didactic, somewhat heavy style. Elton ascribes the "Shield of Hercules" to an imitator of Homer, and in support of this view points out those respects in which the poem resembles, and those in which it is inferior to, the "Shield of Achilles." The two descriptions are, however, absolutely identical in many places; and this would certainly not have happened if one had been an honest imitation of the other. And those parts of the "Shield of Hercules" which have no counterparts in the "Shield of Achilles" are too well conceived and expressed to be ascribed to a very inferior poet—a poet so inferior as to be reduced to the necessity of simply reproducing Homer's words in other parts of the poem. Those parts which admit of comparison—where, for instance, the same objects are described, but in different terms—are certainly inferior in the "Shield of Hercules." The description is injured by the addition of unnecessary or inharmonious details. Elton speaks, accordingly, of these portions as if they were expansions of the corresponding parts of the "Shield of Achilles." This appears to me a mistake. It seems far more likely that both descriptions are by the same poet. It is not necessary for the support of my theory that this poet should be Homer, but I think both descriptions show undoubted traces of his handiwork. Indeed, all known imitations of Homer are so easily recognizable as the work of inferior poets, that I should have thought no doubt could exist on this point, but for the attention which the German theory respecting the "Iliad" has received. Assigning both poems to Homer, the "Shield of Hercules" may be regarded, not as an expansion (in parts) of the "Shield of Achilles," but as an earlier work of Homer's, improved and pruned by his maturer judgment, when he desired to fit it into the plan of the "Iliad." Or rather, each poem may be looked on as an abridgment (the "Shield of Hercules" the earlier) of an independent work on a subject presently to be mentioned.

It is next to be shown that, in the events preceding the "Oplopœia," there is a preparation for the introduction of a separate poem.

In the first place, every reader of Homer is familiar with the fact that the poet constantly makes use, when occasion serves, of expressions, sentences, often even of complete passages, which have been already applied in a corresponding, or occasionally even in a wholly different, relation. The same epithets are repeatedly applied to the same deity or hero. A long message is delivered in the very words which have been already used by the sender of the message. In one well-known instance (in Book II.), not only is a message delivered thus, but the person who has received it repeats it to others in precisely the same terms. In the combat between Hector and Ajax (Book VI.), the flight of Ajax's spear, and the movement by which Hector avoids the missile, are described in six lines, differing only as to proper names from those which had been already used in describing the encounter between Paris and Menelaus (Book III.).

This peculiarity would be a decided blemish in a written poem. Tennyson, indeed, occasionally copies Homer's manner—for instance, in "Enid," he twice repeats the line—

"As careful robins eye the delver's toil;"

but with a good taste which prevents the repetition from becoming offensive. The fact is, that the peculiarity marks Homer as the *singer*, not the *writer*, of poetry. I would not be understood as accepting the theory, according to which the "Iliad" is a mere string of ballads. I imagine that no one who justly appreciates that noble poem would be willing to countenance such a theory. But that the whole poem was sung by Homer at those prolonged festivals which formed a characteristic peculiarity of Achaian manners seems shown, not only by what we learn respecting the later "rhapsodists," but by the internal evidence of the poem itself.*

* Besides Homer's reference, both in the "Iliad" and the "Odyssey," to poetic recitations at festivals, there is the well-known invocation in

Homer, reciting a long and elaborate poem of his own composition, occasionally varying the order of events, or adding new episodes, extemporized as the song proceeded, would exhibit the peculiarity invariably observed in the "improvisator," of using, more than once, expressions, sentences, or passages, which happened to be conveniently applicable. The art of extemporizing depends on the capacity for composing fresh matter while the tongue is engaged in the recital of matter already composed. Any one who has watched a clever improvisator cannot fail to have noticed that, though gesture is aptly wedded to words, the thoughts are elsewhere. In the case, therefore, of an improvisator, or even of a rhapsodist reciting from memory, the occasional recurrence of a well-worn form of words serves as a relief to the strained invention or memory.

We have reason, then, for supposing that if Homer had, in his earlier days, composed a poem which was applicable, with slight alterations, to the story of the "Iliad," he would endeavor, by a suitable arrangement of the plan of his narrative, to introduce the lines whose recital had long since become familiar to him.

Evidence of design in the introduction of the "Shield of Achilles" certainly does not seem wanting.

Book II. To what purpose would the mere writer of poetry pray for an increase of his physical powers? Nothing could be more proper, says Gladstone, if Homer were about to write; nothing less proper if he were engaged on a written poem.

It is by no means necessary to the plot of the "Iliad" that Achilles should lose the celestial armor given to Peleus as a dowry with Thetis. On the contrary, Homer has gone out of his way to render the labors of Vulcan necessary. Patroclus has to be so ingeniously disposed of, that while the armor he had worn is seized by Hector, his body is rescued, as are also the horses and chariot of Achilles.

We have the additional improbability that the armor of the great Achilles should fit the inferior warriors Patroclus and Hector. Indeed, that the armor should fit Hector, or rather, that Hector should fit the armor, the aid of Zeus and Ares has to be called in:

> "To this Jove's sable brows did bow; and he made fit his limbs
> To those great arms, to fill which up the war-god entered him
> Austere and terrible, his joints and every part extends
> With strength and fortitude."—*Chapman's* Translation.

It is clear that the narrative would not have been impaired in any way, while its probability and consistency would have been increased, if Patroclus had fought in his own armor. The death of Patroclus would in any case have been a cause sufficient to arouse the wrath of Achilles against Hector—though certainly the hero's grief for his armor is nearly as poignant as his sorrow for his friend's death.

It appears probable, then, that the description of Achilles's Shield is an interpolation—the poet's own work, however, and brought in by him in the only way he found available. The description clearly refers to the same object which is described (here, also, only in part) in the "Shield of Hercules." The original description, doubtless, included all that is found in both "shields," and probably much more.

What, then, was the object to which the original description applied? An object, I should think, far more important than a warrior's shield. I imagine that any one who should read the description without being aware of its accepted interpretation would consider that the poet was dealing with an important series of religious sculptures, possibly that he was describing the dome of a temple adorned with celestial and terrestrial symbols.

In Egypt there are temples of a vast antiquity, having a dome, on which a zodiac—or, more correctly, a celestial hemisphere—is sculptured with constellation-figures. And we now learn, from ancient Babylonian and Assyrian sculptures, that these Egyptian zodiacs are in all probability merely copies (more or less perfect) of yet more ancient Chaldean zodiacs. One of these Babylonian sculptures is figured in Rawlinson's "Ancient Monarchies." It seems probable that in a country where Sabæanism, or star-worship, was the prevailing form of religion, yet more imposing proportions would be given to such zodiacs than in Egypt.

My theory, then, respecting the Shield of Achilles is this:

I conceive that Homer, in his Eastern travels, visited imposing temples devoted to astronomical observation and star-worship; and that nearly every line in both "shields" is borrowed from a poem in which he described a temple of this sort, its domed zodiac, and those illustrations of the labors of different seasons and of military or judicial procedures which the astrological proclivities of star-worshippers led them to associate with the different constellations.

I think there are arguments of some force to be urged in support of this theory, fanciful as it may seem.

In the first place, it is necessary that the constellations recognized in Homer's time (not necessarily, or probably, *by* Homer) should be distinguished from later inventions.

Aratus, writing long after Homer's date, mentions forty-five constellations. These were probably derived, without exception, from the globe of Eudoxus. Remembering the tendency which astronomers have shown, in all ages, to add to the list of constellations, we may assume that in Homer's time the number was smaller. Probably there were some fifteen northern and ten southern constellations, besides the twelve zodiacal signs. The smaller constellations mentioned by Aratus doubtless formed parts of larger figures.

Any one who studies the heavens will recognize the fact that the larger constellations have been robbed of their just proportions to form the smaller asterisms. Corona Borealis was the right arm of Bootes, Ursa Minor was a wing of Draco (now wingless, and no longer a dragon), and so on.

Secondly, it is necessary that the actual appearance of the heavens, with reference to the position of the pole in Homer's time, should be indicated. For our present purpose, it is not necessary that we should know the exact date at which the most ancient of the zodiac-temples were constructed (or to which they were made to correspond). There are good reasons, though this is not the proper place for dwelling upon them, for supposing that the great epoch of reference among ancient astronomers preceded the Christian era by about 2,200 years. Be this as it may, any epoch between the date named and the probable date at which Homer flourished—say nine or ten centuries before the Christian era—will serve equally well for our present purpose. Now, if the effects of equinoctial precession be traced back to such a date, we are led to notice two singular and not uninteresting circumstances. First, the pole of the heavens fell in the central part of the great constellation Draco; and, secondly, the equator fell along the length of the great sea-serpent, Hydra, in one part of its course, and elsewhere to the north of all the ancient aquatic constellations,* save that one-half of the northernmost fish (of the zodiac pair) lay north of the equator. Thus, if a celestial sphere were constructed with the equator in a horizontal position, the Dragon would be at the summit, Hydra would be extended horizontally along the equator — but with his head and neck reared above that circle — and Argo, Cetus, Capricornus, Piscis Australis, and Pisces — save one-half of the northernmost—would lie *below* the equator. It may, also, be mentioned that all the bird-constellations were then, as now, clustered together not far from the equator—Cygnus (the farthest from the equator) being ten degrees or so nearer to that circle than at present.

Now let us turn to the two "shields," and see whether there is any thing in them to connect them with zodiac-temples, or to remind us of the relations exhibited above. To commence with the "Shield of Achilles," the opening lines inform us that there appeared—

"The starry lights that heaven's high convex crowned,
The Pleiads, Hyads, with the northern team,
And great Orion's more refulgent beam."

And here, in Achilles's shield, the list of constellations closes; but it is remarkable that in the "Shield of Hercules," while the above lines are wanting, we find

* We may exclude Delphinus as probably later than Homer's time, though mentioned by Aratus.

lines which clearly point to other constellations. Remembering what has just been stated about Draco, it seems at the least a singular coincidence that we should find the centre or boss of the shield occupied by a dragon:

> "The scaly horror of a dragon, coiled
> Full in the central field, unspeakable,
> With eyes oblique retorted, that aslant
> Shot gleaming flame."*—*Elton's* Translation.

We seem, also, to find a reference to the above-named relations of the aquatic constellations, and specially to the constellation Pisces:

> "In the midst,
> Full many dolphins chased the fry, and showed
> As though they swam the waters, to and fro
> Darting tumultuous: two † of silver scale
> Panting above the wave."

* Compare the description of the constellation Draco by Aratus:

> "Swol'n is his neck—eyes charged with sparkling fire
> His crested head illume. As if in ire
> To Helice he turns his foaming jaw
> And darts his tongue, barbed with a blazing star."
> *Lamb's* Translation.

† It is scarcely necessary to remark that no importance is to be attached to the numerical relations in this and other passages. In the original work describing a zodiac-dome, the exact number of constellations representing fishes, dogs, or the like, would of course be mentioned; but any changes necessary to Homer's purpose in describing a shield would unhesitatingly have been introduced by him subsequently. It is singular, however, that we should have here, and in the passage quoted farther on as referring to Orion and the Dogs, the number *two*

For we learn from both "shields" that the waves of ocean were figured in a position corresponding with the above-mentioned position of the celestial equator, beneath which—that is, *in the ocean*, on our assumption—the aquatic constellations were figured. The description of the ocean in the "Shield of Hercules" contains also some lines, in which we seem to see a reference to the bird-constellations close above the equator:

> "Rounding the utmost verge the ocean flowed
> As in full swell of waters, and the shield
> All variegated with whole circle bound.
> Swans of high-hovering wing there clamored shrill,
> Who also skimmed the breasted surge with plume
> Innumerous; near them fishes 'midst the waves
> Frolicked in wanton bounds."

In the "Shield of Achilles" no mention is made of Perseus, but in the "Shield of Hercules" this well-known constellation seems described in the lines—

> "There was the knight of fair-haired Danae born,
> Perseus; nor yet the buckler with his feet
> Touched nor yet distant hovered, strange to see,
> For nowhere on the surface of the shield
> He rested; so the crippled artist-god

specially mentioned. The latter instance is the more remarkable, inasmuch as the mention of men and hares would lead one to expect that more than two dogs would be introduced. I would suggest as a sufficient reason for this peculiarity that the verbal alterations necessary to pluralize some of the objects in the dome would be more easily effected than those necessary to undualize others.

Illustrious framed him with his hands in gold.
Bound to his feet were sandals winged; a sword
Of brass, with hilt of sable ebony,
Hung round him from the shoulders by a thong.
. . . . . . . . . The visage grim
Of monstrous Gorgon all his back o'erspread;
. . . . . . . . . the dreadful helm
Of Pluto clasped the temples of the prince."

I think that one may recognize a reference to the twins Castor and Pollux (the wrestler and boxer of mythology) in the words—

"But in another part
Were men who wrestled, or in gymnic fight
Wielded the cestus."

Orion is not mentioned by name in the "Shield of Hercules," as in the other; but Orion, Lepus, and the two dogs, seem referred to:

"Elsewhere men of chase
Were taking the fleet hares; two keen-toothed dogs
Bounded beside, these ardent in pursuit,
Those with like ardor doubling in their flight."

In each "shield" we find a reference to the operations of the year—hunting and pasturing, sowing, ploughing, and harvesting. It is hardly necessary to point out the connection between these operations and astronomical relations. That this connection was fully recognized in ancient times is shown in the "Works and Days" of Hesiod. We find also in Egyptian zodiacs clear evidence that these operations, as well as astronomical symbols or constellations, found a record in sculptured domes.

The judicial, military, and other proceedings described in the "Shield of Achilles" were also supposed by the ancients to have been influenced by the courses of the stars.

If we had no evidence that ancient celestial spheres presented the constellations above referred to, we might be disposed to attach less weight to the coincidences here presented; but the "Phenomena" of Aratus affords sufficient testimony on this point. In the first place, that work is of great antiquity, since Aratus flourished two centuries and a half before the Christian era; but it is well known that Aratus did not describe the results of his own observations. The positions of the constellations, as recorded by him, accord neither with the date at which he wrote nor with the latitude in which he lived. It is generally assumed—chiefly on the authority of Hipparchus—that Aratus borrowed his knowledge of astronomy from the sphere of Eudoxus; but we must go much farther back even than the date of Eudoxus, before we can find any correspondence between the appearance of the heavens and the description given by Aratus. Thus we may very fairly assume that the *origin* of the constellations (as distinguished from their association with certain circles of the celestial sphere) may be placed at a date preceding, perhaps by many generations, that at which Homer flourished.

Indeed, there have not been wanting those who find in the ancient constellations the record of the early history of man. According to their views, Orion is Nimrod—the "Giant," as the Arabic name of the constellation implies—the mighty hunter, as the dogs and hare beside him signify. The Centaur bearing a victim toward the altar is Noah; Argo, the stern of a ship, is the ark, as of old it might be seen on Mount Ararat. Corvus is the crow sent forth by Noah, and the bird is placed on Hydra's back to show that there was no land on which it could set its foot. The figure now called Hercules, but of old Engonasin, or the kneeler, and described by Aratus as "a man doomed to labor," is Adam. His left foot treads on the dragon's head, in token of the saying, "It shall bruise thy head;" and Serpentarius, or the serpent-bearer, is the promised seed.

Of course, if we accept these views, we have no difficulty in understanding that a poet so ancient as Homer should refer to the constellations which still appear upon celestial spheres. And, in any case, the mere question of antiquity presents, as we have already shown, little difficulty.

But there is a difficulty in one respect, a notice of which must close this paper, already carried far beyond the limits I had proposed to myself: It may be thought remarkable that heroes of Greek mythology, as Perseus and Orion, should be placed by Homer, or even by Aratus, in spheres which are undoubtedly of Eastern origin.

Now, it may be remarked, first, of Homer, that many acute critics consider the whole story of the "Iliad" to be, in reality, merely an adaptation of an Eastren narrative to Greek scenes and names. It is pointed out that, whereas the Catalogue in Book II. reckons upward of 100,000 men, only 10,000 fought at Marathon; and, whereas there are counted no less than 1,200 ships in the Catalogue, there were but 271 at Artemisium, and at Salamis but 378. However this may be, we have the distinct evidence of Herodotus that the Greek mythology was derived originally from foreign sources. He says, "All the names of the gods in Greece were brought from Egypt," an opinion in which Diodorus and other eminent authorities concur. But it is the opinion of acute modern critics that we must go beyond Egyptian—to Assyrian, or Indian, perhaps even to Hebrew sources, for the origin of Greek mythology. Bryant traces nearly all the Greek myths to traditions of the dispersion of the Cuthites or Cuscans. And Layard has ascribed to Niebuhr the following significant remarks: "There is a want in Grecian art which neither I, nor any man now alive, can supply. There is not enough in Egypt to account for the peculiar art and the peculiar mythology which we find in Greece.

That the Egyptians did not originate it I am convinced, though neither I, nor any man now alive, can say who were the originators. But the time will come when, on the borders of the Tigris and Euphrates, those who come after me will live to see the origin of Grecian art and Grecian mythology."

(From *The Student*, June, 1868.)

THE END.